This volume reports on the characteristics and biological effects of urbanisation in Third World cities. Several chapters describe the ecology of such cities and other urbanising places, to show exactly which physical and social features of cities may influence human health and biology. Other chapters investigate a wide variety of biological responses to features of urban ecology such as the frequencies of specific diseases, nutritional status, immunological characteristics, precursors of cardiovascular disease, endocrine levels, and patterns of child growth and development.

This important volume will be of interest to a wide range of researchers and academics including human biologists, anthropologists, health care professionals, human geographers, urban and regional planners, and economists.

SOCIETY FOR THE STUDY OF HUMAN BIOLOGY
SYMPOSIUM SERIES: 32

Urban ecology and health in the Third World

PUBLISHED SYMPOSIA OF THE
SOCIETY FOR THE STUDY OF HUMAN BIOLOGY

Numbers 1–9 were published by Pergamon Press, Headington Hill Hall, Headington, Oxford OX3 0BY. Numbers 10–24 were published by Taylor & Francis Ltd, 10–14 Macklin Street, London WC2B 5NF. Further details and prices of back-list numbers are available from the Secretary of the Society for the Study of Human Biology.

Urban Ecology and Health in the Third World

32nd Symposium Volume of the
Society for the Study of Human Biology

EDITED BY

LAWRENCE M. SCHELL
Departments of Anthropology and Epidemiology
University at Albany, State University of New York

MALCOLM T. SMITH
Department of Anthropology
University of Durham

AND

ALAN BILSBOROUGH
Department of Anthropology
University of Durham

 CAMBRIDGE
UNIVERSITY PRESS

CAMBRIDGE UNIVERSITY PRESS
Cambridge, New York, Melbourne, Madrid, Cape Town, Singapore, São Paulo, Delhi

Cambridge University Press
The Edinburgh Building, Cambridge CB2 8RU, UK

Published in the United States of America by Cambridge University Press, New York

www.cambridge.org
Information on this title: www.cambridge.org/9780521103053

© Cambridge University Press 1993

First published 1993
This digitally printed version 2009

A catalogue record for this publication is available from the British Library

Library of Congress Cataloguing in Publication data
Urban ecology and health in the third world / edited by Lawrence M.
Schell, Malcolm T. Smith, and Alan Bilsborough.
    p.   cm. – (Society for the Study of Human Biology symposium
series ; 32)
    Includes index.
    ISBN 0–521–41159–9
    1. Urban health–Developing countries.   2. Urban ecology–
Developing countries.
    I. Schell, Lawrence T.   II. Smith, Malcolm T.
    III. Bilsborough, Alan.   IV. Series.
RA566.7.U73   1993
362.1′0425–dc20   92–14144   CIP

ISBN 978-0-521-41159-2 hardback
ISBN 978-0-521-10305-3 paperback

Contents

Contributors

Dr L. Adair
Carolina Population Center, University of North Carolina, Chapel Hill, NC 27514, U.S.A.

E. Beck
St. Mary's Hospital Medical School, Praed Street, London N2 1PG, U.K.

Dr A. Bilsborough
Department of Anthropology, University of Durham, Durham DH1 3HN, U.K.

Dr A. Bittles
Anatomy and Human Biology Group, Biomedical Sciences Division, King's College, University of London, Strand, London WC2R 2LS, U.K.

Dr A. V. Buchanan
Department of Anthropology, The Pennsylvania State University, University Park, PA 16802, U.S.A.

Dr N. Cameron
Department of Anatomy and Human Biology, Medical School, University of the Witswatersrand, Johannesburg 2193, South Africa.

J. Campbell
Department of Anthropology, University of Oklahoma, Norman, OK 73019, U.S.A.

Professor J. I. Clarke
Department of Geography, University of Durham, Durham DH1 3HN, U.K.

Dr E. Crognier
ER 221 du CNRS, Pavillon de L'enfant, 346, Routes des Alpes, Aix-en-Provence, France.

Professor D. Drakakis-Smith
Department of Geography, University of Keele, Keele, Staffs, U.K.

x   *List of contributors*

Dr W. W. Dressler
Department of Behavioral and Community Medicine, University of
Alabama School of Medicine, Tuscaloosa, Alabama 35487–0326,
U.S.A.

P. Dulieu
Institut Supérieur Pédagogique, Bunia, Zaire.

Dr P. L. Engle
Department of Psychology and Human Development, California
Polytechnic State University, San Luis Obispo, CA 94307, U.S.A.

Dr G. Flannery
Department of Genetics and Human Variation, La Trobe University,
Bundoora, Australia 3083.

Professor J. L. A. Ghesquiere
Institute of Physical Education, K. U. Leuven, Belgium.

Dr T. Harpham
Urban Health Programme, London School of Hygiene and Tropical
Medicine, Keppel Street, London WC1E 7HT, U.K.

Dr C. Jenkins
Papua New Guinea Institute of Medical Research, P.O. Box 60,
Goroka, Eastern Highlands Province, Papua New Guinea.

Dr F. E. Johnston
Department of Anthropology, University of Pennsylvania,
Philadelphia, PA 19104, U.S.A.

Dr H. Kaplowitz
Food Research Institute, Stanford University, Stanford, CA
94305–6084, U.S.A.

E. Katz
Instituto de Investigaciones Antropologicas, U.N.A.M., Ciudad
Universitaria, Del. Coyoacan, Mexico D.F., Mexico.

Dr P. Lefevre-Witier
Centre de Recherches sur le Polymorphisme Genetique des
Populations Humaine, (CRPG), CNRS, Hopital Purpan, Toulouse,
France.

Dr R. Martorell
Food Research Institute, Stanford University, Stanford, CA
94305–6084, U.S.A.

Dr W. M. Mason
Department of Sociology, University of California, Los Angeles, CA
90024, U.S.A.

Dr J. H. Moore
Department of Anthropology, University of Oklahoma, Norman, OK 73019, U.S.A.

E. Nkiama
Institut Superieur Pédagogique, Bunia, Zaire.

Dr J. Pryer
Centre for Human Nutrition, University of Sheffield, Northern General Hospital, Sheffield S5 7AU, U.K.

Dr J. E. Dos Santos
Faculty of Medicine, University of Sao Paulo–Ribeirao Preto, Brazil.

Dr L. M. Schell
Departments of Anthropology and Epidemiology, University at Albany, State University of New York, Albany, NY 12222, U.S.A.

Dr C. Serrano
Instituto de Investigaciones Antropologicas, U.N.A.M., Ciudad Universitaria, Del. Coyoacan, Mexico D.F., Mexico.

S. Shreeniwas
Population Studies Center, University of Michigan, Ann Arbor, MI 48104–2590, U.S.A.

D. N. Singarayer
Anatomy and Human Biology Group, Biomedical Sciences Division, King's College, University of London, Strand, London WC2R 2LS, U.K.

Dr M. Smith
Department of Anthropology, University of Durham, Durham DH1 3HN, U.K.

M. Spinar
Population Studies Center, University of Michigan, Ann Arbor, MI 48104–2590, U.S.A.

Dr R. Valdez
The University of Texas, Health Science Center at San Antonio, San Antonio, TX 78284, U.S.A.

Dr J. VanDerslice
Carolina Population Center, University of North Carolina, Chapel Hill, NC 27514, U.S.A.

Dr L. A. Vargas
Instituto de Investigaciones Antropologicas, U.N.A.M., Ciudad Universitaria, Del. Coyoacan, Mexico D.F., Mexico.

Dr F. E. Viteri
Department of Nutrition, University of California-Berkeley, Berkeley, California 94720, U.S.A.

Dr K. Weiss
Department of Anthropology, The Pennsylvania State University, University Park, PA 16802, U.S.A.

R. Wellens
Institute of Physical Education, K. U. Leuven, Belgium.

Dr N. White
Department of Genetics and Human Variation, La Trobe University, Bundoora, Australia 3083.

Dr C. Worthman
Department of Anthropology, Emory University, Atlanta, GA 30322, U.S.A.

Dr B. Zemel
Nutrition and Growth Laboratory, The Children's Hospital of Philadelphia, Philadelphia, PA 19104-4399, U.S.A.

Dr N. Zohoori
Carolina Population Center, University of North Carolina, Chapel Hill, NC 27514, U.S.A.

Acknowledgements

The editors wish to express their appreciation to Elizabeth Watts who helped organise the symposium on which this volume is based, and to acknowledge the support of St. Mary's College of Durham University, and of the College of Social and Behavioral Sciences of the University at Albany, State University of New York.

1 Human biological approaches to the study of Third World urbanism

L. M. SCHELL, M. T. SMITH AND A. BILSBOROUGH

Of the world's population in 1991 of 5384 million, it is estimated that 43% live in urban areas (Population Reference Bureau, 1991). Current projections based on data assembled by the United Nations predict that one half of the world's population will be living in urban areas by the year 2005 (United Nations, 1989).

This trend is a continuation of more than several thousand years of urbanisation. However, in the past two centuries the rate of urbanisation has accelerated. Between 1800 and 2000, the world's urban population is expected to have increased 128-fold, yet the world's total population will have increased but 6.4-fold over the same period (Rogers & Williamson, 1982).

Urbanism is unequally distributed among the more and less developed nations. For more than ten years, the urban proportion of the population in developed nations has exceeded 70% whereas among the less developed nations the percentage of urban residents has remained below 35% (United Nations, 1989). Although the less developed nations are now less urban, their rate of urban growth is greater (Kurian, 1991). All of the eleven nation-states with the highest annual rates of urban population growth are classed as less developed nations: Sri Lanka, Zaïre, Tanzania, Bangladesh, Oman, Yemen (North), Cameroon, Niger, Ivory Coast, Libya and Rwanda (Kurian, 1991). According to United Nations projections (medium variant model; 1989), by the year 2015, fully half of the population in the less developed nations will reside in urban areas, and these nations will have an urban population more than three times that of the more developed nations. Mexico City alone is expected to have a population of 31 million (Kurian, 1991).

There are few changes in the history of human existence comparable to urbanisation in scope and potential to bring about biologic change. Only the transition from a society based on hunting and gathering to one based on agriculture seems its equal. The consequences of that change are now being understood; scholars have attributed some of the most common chronic diseases in populations of developed nations to dietary changes

1

accompanying the transition from hunting and gathering subsistence to one based on agriculture (Eaton and Konner, 1985). The second transition in Europe, from an agricultural to an industrial urban society, has already produced substantial changes in human health (Wohl, 1983), morphology and growth (Tanner, 1986). Future social changes associated with continued urban growth will produce more biologic change as well.

Just what changes to expect in the future, especially in the less developed nations, is not clear. The European experience may not provide an appropriate model for predicting change in the less developed nations. European urbanisation has taken over 200 years to occur, and has been driven largely by indigenous forces such as industrialisation, unfolding continuously over the entire period. In the less developed nations industrialisation is imported in its current form, and its growth is usually encouraged as a means to economic development. As an import which is rapidly growing, industrialisation may entail more socially and ecologically disruptive changes in the less developed nations with greater consequences for human biology than occurred in Europe.

The rapid urbanisation of the less developed nations raises two interrelated questions: what characterizes the environment of Third World cities, and what is the effect of this environment on human health and on human biology generally?

The approach to urbanism from human biology

Any answer to these questions from the perspective of human biology necessarily involves particular research areas. Human biology traditionally includes studies of genetic characteristics of populations, population structure, demographic parameters such as infant mortality and fertility, morphologic characteristics of the adult population, patterns of physical growth and ageing, and the physiological parameters representing the responses or adaptations of organ systems to environmental stressors.

The approach used by human biologists is well represented by human biological research on such environmental stressors as extremes of temperature and altitude. Here, human biology research has a long history, and many of the focal questions of the discipline can be identified clearly: what is the contribution of the environmental stressor to human variability, what are the individual differences in response to these stressors, and what are the genetic differences or population differences in response to these stressors? Current research in urban ecology and biology may also follow the human biology paradigm. Important unifying questions concern the contribution of urban conditions to human biological variability, and the extent of variation in response to urbanism. Does urbanism produce biological reponses that contribute to variability among populations? Is there biological variation in response to urban stressors, both at the level of individual differences and at the level of variation among populations?

The study of the city as an environment, to which people respond biologically, necessarily involves other questions and also another type of analysis. Because cities are more than a physical environment with certain characteristics of temperature, humidity, noise, etc., studying the urban ecosystem also calls for analysis and measurement of the social context, especially its biologically salient sociocultural characteristics. Although consideration of sociocultural variables has characterized the study of human biology for many years, these variables are more prominent features of the urban environment simply because that environment is entirely human-made and many physical stressors of the natural environment are bufferred. Thus, the study of Third World urban ecology and health necessarily involves detailed analyses of the biosocial interaction as a means to pursue the fundamental questions characterising human biology inquiry.

Research agenda for a new discipline

Although the influence of urban environments on human morphology, health and behaviour has been a subject for commentary by classical writers, sixteenth century social philosophers and nineteenth century natural historians, the scientific study of biological responses to urban environments is a recent development. Contemporary works by human biologists are few; Harrison and Gibson's volume *Man in Urban Environments* was published in 1976. Clegg and Garlick edited *Disease and Urbanisation* in 1980, which focused intensely on disease within the larger arena of human biology. There has not been a comprehensive or representative sampling of research on the human biology of urbanism since these seminal works, despite the large number of human biologists now studying urban populations.

Given the short history of scientific work in this research area and the complexity of analysing both biological and social variables, several research activities are appropriate. These include experimentation with research designs, and the evolution of a set of standard definitions and methods. These activities are clearly represented in current works on urban human biology.

The research design that has been employed most frequently is the urban–rural comparison. In fact, the urban–rural contrast has been a traditional research design not only in specifically urban studies, but also in many areas of human biology (e.g. in studies of children's physical growth). The design is straightforward but has relied upon the discipline's inexact knowledge of the influences within the urban ecosystem. So long as the actual urban features influencing growth patterns, for example, were unknown and urban environments shared many characteristics in common, it was possible to generalise across studies and generate a composite picture of child growth in cities. With the accumulation of results using this design, inconsistencies arose which could be explained only by viewing the urban environment not as a constant, but as complex and highly variable. Whether there now exists more variability among cities, or whether we more clearly perceive variability between and within cities, few researchers still believe that all cities have a similar effect on human growth, to choose but one example. (See Bielicki, 1986, for a cogent analysis of urban–rural growth comparisons.)

Another trend within the discipline which contributes to the research agenda for urban human biology is the greater sensitivity to methodologic issues in the study of large, contemporary populations. Epidemiologists have pointed out that the main problem with a comparison of two naturally occurring groups, such as an urban and a rural settlement, is that there are likely to be many differences between them, and no one difference can be considered the cause of the others. This is a specific instance of the ecological fallacy in which one feature of the environment, usually the only measured one, is held responsible for the biological outcomes differing between the populations even though other factors were not measured and their influence not tested (Morganstern, 1982).

The simple urban–rural comparison is usually open to this criticism. For example, an urban sample and a rural sample of children may differ in their average blood lead levels and in their physical growth, but there are likely to be so many other differences additional to the level of lead in the blood that attributing the growth contrasts to the one difference that is actually measured cannot be justified.

Analysing the urban environment

Current work is moving away from research designs based on the urban–rural comparison with its simplifying assumptions. An important step taken by many researchers is towards analysing the urban environment and determining its constituent features. This is a necessary step for the direct measurement of urbanism, and eventually the specification of its components as independent variables.

One feature of mature areas of scientific research is the presence of a standard method of measuring the independent variable of interest (for example, hypoxia, or temperature). In the history of the natural sciences, being able to measure the phenomenon of interest is basic to advances in understanding its properties and interrelationships. That stage has not yet been reached in many areas of human biology, and currently there is no universally accepted definition of 'urban'.

In the natural sciences a uniform definition is usually achieved after a long process of experimentation in which different ways of measuring the independent variable of interest are tried. For instance, in studies of morphological variation among human populations in relation to temperature (see, for example, Roberts, 1978; Steegman, 1979) there are several measurement scales of temperature that have been employed, and the most common, mean annual temperature, is but one.

Much work on the human biology of urban environments pertains to the issue of what to measure and how to measure it. The urban environment is densely packed, and its constituents differ in different nations and societies. It is this variation that drives the need to break down the complex ecosystem into measurable components. Eventually, these could be measured with a precision equal to that obtained in measuring the biological outcomes which are the focus of human biology inquiry.

A symposium on urban ecology and health in the Third World

The conference at Durham in 1991 was held to explore questions of Third World ecology and human biology by bringing together specialists working on a wide range of topics in a variety of geographic areas. The organisers sought to take a cross-sectional view of the discipline's recent research and of its potential for future work on the impact of urban population growth upon human biology.

A sampling from every area of current research is impossible owing to the great breadth of approaches and subjects within the range of human biology. However, the specific inquiries offered by the contributors to this volume represent many of the traditional subject areas of human biology. Studies of growth and urbanism are contributed by Adair and colleagues, Beck, Cameron, Ghesquiere and colleagues, Johnston, and Pryer. Some of these focus on the relationship of physical growth to nutritional status and some explore relationships to household characteristics such as sanitation, production resources, and socioeconomic status generally. Studies of population-level phenomena include work by Weiss and colleagues on disease prevalence among Native American communities, and by Flannery and White on immunological parameters, while Bittles and colleagues investigate the relationship of urbanism to sex ratio at several levels of population aggregation. Physiological parameters are the subject of reports by Dressler and colleagues, who focus on serum lipid levels which are pertinent to the development of cardiovascular disease, and by Zemel, Jenkins and Worthman, who investigate the growth-regulating endocrines.

Other studies focus on urban ecology itself rather than on biologic responses to urban environments. Drakakis-Smith describes the means of food acquisition for inhabitants of Third World cities. Crognier analyses the social, cultural and biological characteristics of individuals in the villages, towns and cities of Marrakesh province to determine which ones vary with the urban–rural continuum and can be used as a metric for urbanism. The forces pushing migrants from rural settlements and those pulling them to large cities are the focus of analyses by Kaplowitz and Lefevre-Witier and their collaborators. Clarke provides a continent-wide comparison of urbanism and urbanisation in Africa, and Harpham examines urban policy formation in Third World nations.

Two methodological advances in the study of urban human biology are represented among many of the contributions to this volume. Firstly, researchers are experimenting with different conceptualisations of urban. The urban–rural polarity, which has been a convenient fiction for decades, has been replaced by the concept of an urban–rural continuum. Now researchers focus on locations along the urban–rural continuum, and provide sufficient description of the urban character of their populations to permit other workers to locate these populations on the continuum and eventually to generalise across studies.

Thus Zemel and colleagues compare endocrine levels between two settlements, one with many rural characteristics, the other with a larger and denser population, more cosmopolitan diet and educational institutions. Neither population exactly fits the rural or urban archetypes. In the study of migration patterns by Kaplowitz *et al.*, several villages are compared, none of which is decidedly more urban than the others. Each is described in great detail and the analysis of these details enables considerable understanding of the forces driving migration to large urban centres such as Guatemala City. Likewise, Beck compares several Australian aboriginal settlements, termed 'town camps', in his study of nutritional status, disease and sociocultural correlates of urbanism. Cameron also abandons the urban–rural dichotomy in favour of a comparison of children from two rural areas that differ in measurable ways, and a further comparison of children from different socioeconomic groups within one city. Weiss *et al.* compare several groups differing in the degree of modernity in terms of specific diseases and the symptom groups usually antecedent to them.

The second methodologic advance is the precise measurement of variables representing the components of urban ecology at a finer level than the settlement. In several studies the unit of comparison is not the settlement, such as town camps, but households or individuals. Although it has long been common to measure physiological, morphological and genetic characteristics of individuals, measurement of the social components of urbanism, when done at all, usually has been done at the level of the settlement, so leading to comparisons of particular settlements. To understand the effect of urban features rather than of particular settlements, the sociocultural characteristics constituting urbanism are measured at the same level as the particular biological outcomes that are favoured subjects for human biologists. It then becomes possible to examine the relationship of these biological outcomes to particular features of urbanism, with generalisations about urbanism and its constituent features as the goal.

Several studies within the present volume measure urban features at the level of the individual or the household. Pryer reports on an especially detailed analysis using the household as the unit of study. The household is the locus of production and distribution of resources pertaining to nutritional status. Nutritional status is measured in individuals and compared across different types of households as distinguished by cluster analysis. Dressler and collaborators measure psychosocial stress, one of the primary characteristics of urban ecosystems, in individuals to determine its relationship to serum lipid levels. Johnston produces a fine-grained analysis of children's size, cognitive development and school achievement, all of which are measured in individuals. Likewise, Adair focuses on characteristics of communities, households and individuals to determine their relationships to health outcomes, and to one another. The patterns of their interrelationship in urban and rural areas are compared. The approach includes the specific measurement of the microenvironment in urban and rural households and provides a very detailed analysis of the influence of specific features of urban environments on the physical growth of infants.

One other noticeable feature of the contributions to this volume is the variety of cities examined. Among those represented are Soweto, South Africa; Cebu City, Philippines; Harare, Zimbabwe; Bunia, Zaïre; Guatemala City, Guatemala; Khulna, Bangladesh; and numerous lesser known villages and towns that are urbanising, such as the town camps of Australian aboriginal populations. The variety of Third World cities which are the subject of this volume reflects an important trend in the current phase of urban human biology studies: the effort to obtain a wide base of observations about cities in order to move from particular case studies towards broader generalisation, theory construction, and finally specific hypothesis testing.

The organisation of this volume is based on the level of analysis or comparison employed in each contribution. Analyses involving measurement of individuals and comparison of groups smaller than the settlement are listed before descriptions or comparisons of settlements, nations and continents. Since many studies measure different variables at different levels (individual, household, settlement, etc.), slightly different orderings of these studies are equally possible. Another logical organisation would be by subject, and to some degree subject groupings correspond to the ordering by level of measurement and comparison. One group includes studies of physiological parameters measured on the individual level. Comparisons of growth patterns among settlements form another cluster, as do studies of population characteristics such as sex ratio or admixture. Another group focuses on urban ecology, including the subsistence system operating in Third World cities and the forces driving migration. Finally, there are the large scale comparisons such as those by Clark and Harpham.

Current research has moved beyond the initial question of the effect of the urban environment. As previous researchers sought to answer this question they learned that the urban environment was not a unitary phenomenon. As we distinguish the components of the urban environment, we become able to ask specific questions about those components. Instead of the unanswerable question, 'what is the effect of the urban environment?', we now ask 'what is the effect of each factor at a specific level of intensity or dose, alone or in combination with other factors?' The benefit of this development is likely to be an increased ability to generalise about human biological response to ecological change, and to predict more precisely the biological, medical and other outcomes of the seemingly irreversible and accelerating trend towards urbanism.

References

Bielicki, T. (1986). Physical growth as a measure of the economic well-being of populations: The twentieth century. In *Human Growth*, vol. III. *Methodology, Ecology, Genetic, and Nutritional Effects on Growth* (2nd edn) (ed. F. Falkner & J. M. Tanner), pp. 283–305. New York: Plenum Press.

Clegg, E. J. & Garlick, J. P. (1980). *Disease and Urbanisation*. London: Taylor & Francis.

Eaton, S. B. & Konner, M. (1985). Paleolithic: A consideration of its nature and current implications. *New England Journal of Medicine* 312, 283–9.

Harrison, G. A. & Gibson, J. B. (eds) (1976) *Man in Urban Environments*. Oxford University Press.

Kurian, G. T. (1991). *The New World Book of World Rankings* (3rd edn). New York: Facts on File.

Morganstern, H. (1982). The uses of ecological analysis in epidemiologic research. *American Journal of Public Health* 72, 1336–44.

Population Reference Bureau (1991). *1991 World Population Data Sheet*. Washington, D.C.: Population Reference Bureau.

Roberts, D. F. (1978). *Climate and Human Variability* (2nd edn). Menlo Park, California: Cummings Publishing.

Rogers, A. & Williamson, J. C. (1982). Migration, urbanisation, and third world development: an overview. *Economic Development and Culture Change* **30**, 463–82.

Steegman, Jr., A. T. (1979). Human adaptation to cold. In *Physiological Anthropology* (ed. A. Damon), pp. 130–66. New York: Oxford University Press.

Tanner, J. M. (1986). Growth as a mirror of the condition of society. In *Human Growth: A Multidisciplinary Review* (ed. A. Demirjian), pp. 3–34. Basingstoke, Hants: Taylor & Francis.

United Nations (1989). *World Population Prospects 1988*. New York: United Nations.

Wohl, A. S. (1983). *Endangered Lives. Public Health in Victorian Britain*. Cambridge, Massachusetts: Harvard University Press.

2 Social and cultural influences in the risk of cardiovascular disease in urban Brazil

W. W. DRESSLER, J. E. DOS SANTOS AND F. E. VITERI

Throughout the twentieth century there has been a marked acceleration of processes of modernisation and urbanisation in developing societies. Cities in the Third World, and especially urban centres of South America, have swollen with an influx of migrants from the countryside. In many cases these migrants have been driven by failing rural economies and natural disasters, but of equal importance is the allure of city life. The urban centre holds out the promise of 'the good life'; it offers the possibility of achieving the kind of lifestyle and affluence symbolised in media imported from the industrial centres of Europe and North America.

Concomitantly, developing societies have been progressing through that transformation of patterns of morbidity and mortality known as 'the epidemiologic transition'. Rates of infectious and parasitic disease have remained high, but rates of chronic disease, and especially coronary heart disease (CHD), have climbed at an alarming rate. For example, in urban Brazil in 1930, infectious and parasitic diseases accounted for half of all deaths, while CHD accounted for only 12%. By 1980, infectious and parasitic diseases accounted for 12% of all deaths, while CHD accounted for 33% (James et al., 1991).

How are we to account for this transition? Conventional wisdom would argue that as modernisation and urbanisation proceed, life becomes more sedentary and diets become more sodium- and fat-laden, which in turn increases blood pressure and unfavourable lipid profiles, and leads to CHD. This is a hypothesis of considerable merit, but it is also unduly limited, ignoring the host of social and behavioural changes that accompany modernisation and urbanisation, changes that alter individual behaviours and influence physiologic adjustment. The aim of this paper is to examine these issues in detail, and to test specific hypotheses using data collected in an urban area in southwestern Brazil.

10

Social change and physiologic adjustment

Rates of CHD increase within a society as modernisation proceeds; these rates are mediated by increasing population average blood pressures and serum lipids (Dressler, 1984). This association has usually been examined by using a unidimensional model of modernisation; that is, societies are viewed as slowly changing in the direction of industrial society, which means that 'modern' traits get added in to communities, or on to individual behaviour. These traits include wage-labour occupations, formal education, a non-kin pattern of social interaction, adoption of the English language (or some other cosmopolitan language), and formal literacy, along with other factors. Whether the accretion of these traits is measured for individuals or communities, their accumulation is associated with higher blood pressure (Dressler, 1984) and with an unfavourable pattern of serum lipids, including higher total cholesterol (TC) and lower high-density lipoprotein cholesterol (HDL-C) (Barnicot *et al.*, 1972; Labarthe *et al.*, 1973; Page *et al.*, 1974; Pelletier & Hornick, 1986). It is precisely this pattern of risk factors, along with a higher ratio of TC to HDL-C, that is associated with the risk of CHD in industrial societies (Grundy, 1986).

As useful as these studies have been, each is predicated on an outmoded view of the process of social change or modernisation. It is clear that Third World societies have not gradually evolved into industrial societies. Some societies have been able to develop substantial industrial sectors of their economies, but at the expense of staggering foreign debt and hence an enforced dependency on developed societies. Social change or modernisation must be viewed in this context of interlocking dependency in world economic systems, and the search for microlevel social and behavioural changes that may influence physiologic adjustment must also be conducted with this in mind.

Dressler (1982, 1990; Dressler *et al.*, 1986*a,b*, 1987*a,b*) developed a model to examine blood pressure in developing societies which assumes these more complex macro- and microsocial interactions. A fundamental concept in this model is 'style of life', defined as the accumulation of material goods and the adoption of behaviours intended to symbolise high social status or prestige. Anthropological research has consistently shown in studies in the Third World that the lifestyle that comes to be valued in this sense is that characteristic of Euroamerican middle classes, as this is depicted in the mass media. The image of 'the good life', and hence a life that encapsulates prestige, is one of material consumption. By itself, this is not particularly problematic; however, while the media images in television, the cinema, and magazines depicting this lifestyle quickly diffuse through a society, the actual expansion of the economy to support these lifestyle aspirations is much slower. Slow rates of upward economic mobility and high rates of unemployment ensure that many persons will aspire to a high-status lifestyle but be unable to achieve a place in the occupational class hierarchy commensurate with those aspirations. This discrepancy, referred to as 'lifestyle incongruity', is associated with higher blood pressure in several different contexts of social change, after the better-known correlates of blood pressure (including diet) have been controlled (Dressler, 1992).

In essence, lifestyle incongruity locates an individual in a social space fraught with contradictions. He is projecting a sense of himself as a high-status, thoroughly 'modern' person, but he in fact ranks low in the more objective measure of status (occupation). It is therefore unlikely that he will be treated with the respect he desires based on his lifestyle aspirations, and this is likely to lead to a state of chronic psychophysiologic arousal. There are, however, other social factors that modify this chronic arousal. Kinship and other social networks, along with lifestyles, are important determinants of an individual's position in social space. Through being embedded in networks of kin and other social relationships, the individual can develop a sense of his importance in the social world, a sense which may moderate or 'buffer' the contradictions of lifestyle incongruity. In research on blood pressure, these kinds of social supports have been found to interact with lifestyle incongruity in the prediction of blood pressure, such that the deleterious effect of lifestyle incongruity is moderated among those who perceive themselves to have greater social support (Dressler, 1990).

The deleterious effects of high lifestyle incongruity and low social support may prove especially problematic in the urban environment in the Third World. Urban centres in the Third World have grown at a geometric rate in this century, and especially in the past four decades. The major cities of the developing world offer the migrant the hope of achieving a lifestyle which he has seen in books and magazines and on the television. This is both because of the greater availability of goods in the urban centre, and because of perceived economic opportunities there. This promise is of course compromised by the slow economic expansion and high unemployment noted above, and the physiologic adjustment of the migrant can further be compromised by the disruption of traditional social support systems. Very often the migrant will be leaving behind his main sources of support, even if he is using a kin-reliant adaptive strategy (Graves & Graves, 1980) in the process of migrating. Social adaptation in this sense can then require the formation of a non-kin system of social support, a strategy which may or may not be effective (Dressler, 1992).

As noted, this model has been replicated in a variety of different settings using blood pressure as a dependent variable. These are good reasons to suppose that this model may also prove useful in the study of serum lipids.

Correlates of serum lipids

While the risk of CHD associated with elevated serum lipids has been established without question, the factors associated with varying levels of lipids are less well understood. Aside from the specific familial forms of hyperlipidaemia, dietary intake of cholesterol and saturated fats has been suspected as a precursor of an unfavourable profile of serum lipids, based primarily on metabolic ward studies (Grundy, 1986). Outside these tightly controlled, and hence rather artificial, conditions, few studies have demonstrated an association between dietary fats and serum lipids. This paucity of evidence has been attributed to the difficulties in accurately estimating dietary intake by using dietary recall techniques (Keys, 1988).

In addition to methodological issues, it may be that there has been insufficient attention paid to the variety of factors that might influence lipids. Several research reports suggest that individuals who have difficulty in social relationships also have higher TC and/or lower HDL-C (Thomas *et al.*, 1985; Weidner *et al.*, 1987). In Third World sociocultural contexts, then, the model developed for blood pressure may apply equally well to serum lipids. In the remainder of this paper, this model will be examined using data from urban Brazil.

Social change in Brazil

In some senses, Brazil has been a model of successful modernisation. Brazil has developed a potent industrial sector and produces many goods for export. This has, however, been at the expense of a heavy foreign debt, marked inflation, and widespread unemployment. The combination of declining rural economies and the growth of industry in the southwest of the country has led to a steady stream of migrants to the major and provincial urban centres of Sao Paulo and other states. This migration has been exacerbated by years of drought in the northeast.

The research site for the study reported here is Ribeirao Preto, a city of nearly half a million persons in the state of Sao Paulo. Just after the second world war Ribeirao Preto was almost one fifth of its current size. It developed in the late nineteenth century as a marketing and service centre for plantations in this rich agricultural region. In recent decades its growth has been stimulated by the development of industry and financial services, as well as by agricultural innovations. Many of the plantations in the region have been converted to the cultivation of sugar cane, a significant proportion of which is distilled into alcohol fuel for automobiles. The need for unskilled labour in the fields, as well as the promise of higher-status occupations in other economic sectors, has led to population growth.

The emphasis on changing styles of life in Ribeirao Preto is readily apparent in the array of goods available in local shops, and especially in the messages in television and the cinema. These media have been particularly important in Brazil, which has a major domestic television industry. The so-called 'soap opera' has been a staple of Brazilian television, with the depiction of Brazilian middle- and lower-middle-class life, '... where the dominant themes tended to be upward mobility, consumption, and elegant life-styles' (Straubhaar, 1989, p. 239). These themes are repeatedly emphasised as the appropriate aspirations in life.

With respect to social support systems, like many South American cultures with Iberian origins, Brazilian culture has emphasised a strongly patrifocal family. Networks of extended kin have been the traditional source of material, instrumental, and emotional support, and these kin networks have carried over in adaptation to the demands of urban life. At the same time, social change, and especially the changing status of women, has led to a new emphasis on non-kin sources of social support. For migrants to cities, successful adaptation often requires that non-kin systems of social support be established. For men this will often happen in the workplace. For women, this will often happen in the neighbourhood, although with increasing numbers of women entering the workforce, non-kin support systems organised within occupational settings are also important.

In this setting, the effects of lifestyle incongruity, social support and the interaction of these variables on serum lipids will be examined relative to the effects of anthropometric and dietary variables.

Methods

Sampling

The range of socioeconomic diversity in the city of Ribeirao Preto was sampled by using a two-stage design. In the initial step, four residential areas that sampled groups differing in their economic-sector participation were identified. The first, referred to as *boias frias* (literally 'cold plates'), are unstably employed sugar cane cutters, many of whom have migrated from the northeast. The *boias frias* generally live in *favelas*. These are squatter settlements on the outskirts of the city. *Boias frias* cannot be certain of employment from day to day, and other forms of unskilled labour often fill the gap. The second group are agricultural labourers employed full-time on a plantation just outside the city. Workers are continuously employed, with housing and their children's education provided on the plantation. This group is also internally differentiated by occupation: some individuals are unskilled workers, while others are technicians and managers.

Factory workers at a local plant producing audiovisual equipment make up the third group. These workers live on the outskirts of the city in a factory-owned housing development. House plots are purchased from the factory, from which low-interest loans are received for house construction. This group is also internally differentiated. The fourth and final group are bank employees, ranging from clerks to managers, from one of the larger financial concerns in the city. These persons live in traditionally middle- and upper-middle-class neighbourhoods.

Within each of these economic sector or residential groups twenty households were randomly selected for study; in the present analysis data are included from head of household and spouse (where present). Complete data are available for 116 individuals.

Measurement of serum lipids

Blood samples and other medical data were obtained in health centres, either in the community (*boias frias* and agricultural workers) or in the workplace (factory and bank workers). Nurses drew blood samples, took blood pressures, and measured height and weight according to a standard protocol. All blood samples were 12–16 h fasting samples. Standard techniques were used to analyse for total cholesterol (TC) and high-density lipoprotein cholesterol (HDL-C) (Bucolo & David, 1973; Allain *et al.*, 1974; Kostner, 1976). TC values were obtained with a Technicon Autoanalyzer II. HDL-C determinations were carried out by using dextran sulphate to precipitate VLDL and LDL fractions. In addition to using direct values of TC and HDL-C as dependent variables, the TC:HDL-C ratio will also be examined.

Dietary intake

Dietary intake was estimated for each individual as an average of four 24 h dietary recalls. These recalls, collected by trained nutritionists, were converted to nutrient intake using the United States Department of Agriculture food tables, supplemented with Brazilian foods. In previous research (Grundy, 1986), the specific intake of different types of fat has been found to have different effects on lipid profiles. In a detailed analysis of the dietary data from Brazil, we found different kinds of fat in the diet to have similar effects on lipids. Therefore, in the present analysis, we will use total fat intake as a dietary variable. Also, because of their implications in this process, the effects of dietary cholesterol and fibre will be examined.

Sociocultural variables

Lifestyle incongruity was calculated from two component scores: a scale of style of life and a measure of household occupational class. The scale of style of life consists of reported ownership of a number of material goods, such as a radio, television, cassette tape player, automobile, house, and others, as well as reported behaviours regarding exposure to mass media (i.e. reading books and magazines, watching television) and travel. Household occupational class was measured as the sum of occupational rankings of all employed persons in the household; the occupational ranking measures the prestige associated with occupations and is highly correlated with income and education. Lifestyle incongruity is calculated by subtracting occupational class from style of life, after the two scales have been standardized to the same metric. The incongruity score is normally distributed about a mean near zero. (See Dressler *et al.* (1987*b*) for a more detailed discussion.)

As discussed elsewhere (Dressler *et al.*, 1987*a*), testing the incongruity hypothesis requires that the sum of the effects of style of life and occupational class be taken into account, as well as the difference. Here, the sum of style of life and occupational class, which assesses overall vertical ranking in the social class structure, will be termed 'socioeconomic rank'.

Social support was assessed by asking respondents to whom they would turn for help in response to a variety of common problems. They were asked specifically if they would seek aid for each of five problems from a relative, a friend, a neighbour, or a *compadre*. Factor analysis (not shown) of these data demonstrated that responses could be collapsed to two dimensions of social support: kin support and non-kin support. Total scores were calculated for each dimension.

Covariates

Covariates include age (in years), sex (male = 1; female = 0), and the Quetelet Index of body mass, which equals weight (in kilograms) divided by height (in metres) squared.

Results

Descriptive statistics for all variables are shown in Table 2.1. Data have been pooled for males and females, because there were few sex differences.

Table 2.1. *Descriptive statistics for serum lipids, covariate dietary variables, and sociocultural variables* (n = 116)

Variable	Mean (± s.d.)
Total cholesterol	182.4 (± 38.7)
High-density lipoprotein cholesterol	45.9 (± 10.9)
TC:HDL-C ratio	4.2 (± 1.2)
Age	33.6 (± 10.6)
Sex (% male)	44.8
Body mass index	24.1 (± 3.9)
Dietary cholesterol (mg)	245.1 (±163.9)
Total fat (g)	99.5 (± 41.1)
Fibre (g)	8.9 (± 4.9)
Socioeconomic rank	101.2 (± 18.3)
Lifestyle incongruity	0.7 (± 7.2)
Kin support	2.7 (± 1.5)
Non-kin support	2.2 (± 1.7)

Men have significantly higher body mass indices, as well as higher intakes of cholesterol, fat, and fibre. They also perceive lower kin support. With these exceptions there are no sex differences. The amount of total fat consumed is 42.7% of total calories.

Multiple regression analysis was used to examine the relative effects of covariates, dietary variables, and sociocultural variables in relation to TC, HDL-C, and the TC:HDL-C ratio. In relation to each dependent variable, all variables, including interaction effects between lifestyle incongruity and each of the social support variables, were entered concurrently into the analysis, and then non-significant ($p > 0.10$, 2-tailed) effects were deleted on a stepwise basis, in order to arrive at parsimonious models. Non-significant effects were retained only if they were necessary to the valid estimation of an interaction effect. Use of this definition of statistical significance ($p = 0.10$, 2-tailed) is most appropriate here for two reasons. First, all of the hypotheses embedded in the model to be tested are directional, so that this definition of significance corresponds to one-tailed tests at the conventional level of significance. Second, this is a relatively low-power research design given the modest sample size. In this context, a more stringent inclusion criterion would be inappropriate both for evaluating relationships of theoretical interest and for retaining covariates (Rosnow & Rosenthal, 1989).

These results are shown in Tables 2.2–2.4 for each dependent variable. For TC, the final model reduces to the main effects of age, the body mass index, dietary cholesterol, socioeconomic rank, and sex (Table 2.2).

A considerably more complex model is retained for HDL-C (Table 2.3). Higher HDL-C is associated with higher dietary fibre and cholesterol, and

Table 2.2. *Regression of total cholesterol (TC) on predictor variables* *(n = 116)*

(For final step of backward stepwise regression, see text.)

Variable	Standardised regression coefficient
Age	0.372***
Sex	−0.154*
Body mass index	0.251***
Dietary cholesterol	0.188*
Socioeconomic rank	0.164*
$R=$	0.556***
$R=$	0.309

Significance levels: *$p <0.10$, **$p <0.05$, ***$p <0.01$.

Table 2.3. *Regression of high-density lipoprotein cholesterol (HDL-C) on predictor variables (n = 116)*

(For final step of backward stepwise regression, see text.)

Variable	Standardised regression coefficient
Dietary cholesterol	0.214**
Fibre	0.362***
Total fat	−0.353***
Socioeconomic rank	0.253***
Lifestyle incongruity	−0.162*
Kin support	−0.018
Non-kin support	−0.327***
Incongruity × kin support	0.261*
Incongruity × non-kin support	0.343**
$R=$	0.584***
$R=$	0.341

Significance levels: as in Table 2.2.

lower fat intake. Socioeconomic rank is associated with higher HDL-C, and higher lifestyle incongruity with lower HDL-C. The effect of lifestyle incongruity is, however, moderated by kin support and non-kin support, as shown by the significance of the interaction effects. The pattern of these effects is illustrated in Fig. 2.1 for kin support and Fig. 2.2 for non-kin support. The association of low HDL-C with high lifestyle incongruity is concentrated among persons with low social support from either kin or non-kin.

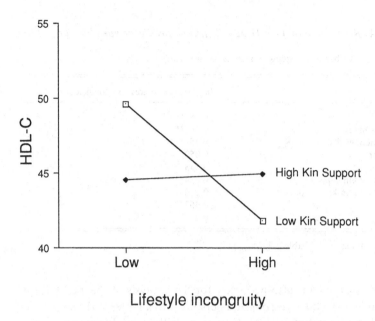

Fig. 2.1 HDL-C in relation to lifestyle incongruity by kin support.

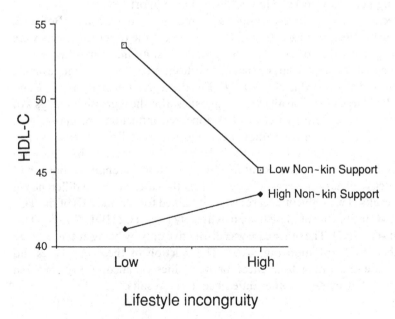

Fig. 2.2 HDL-C in relation to lifestyle incongruity by non-kin support.

Table 2.4. *Regression of TC : HDL-C ratio on predictor variables* (n = 116)

(For final step of backward stepwise regression, see text.)

Variable	Standardised regression coefficient
Age	0.184**
Body mass index	0.196**
Lifestyle incongruity	0.251***
Kin support	0.050
Non-kin support	0.133
Incongruity × kin support	−0.284**
Incongruity × non-kin support	−0.200
$R =$	0.467***
$R =$	0.218

Significance levels: as in Table 2.2.

A complex model is also necessary for the analysis of the TC : HDL-C ratio (Table 2.4). Older persons and persons with a larger body mass index have higher TC relative to HDL-C. Persons with higher lifestyle incongruity also have a higher TC : HDL-C ratio; but again, this is concentrated among persons who have low levels of kin support (see Fig. 2.3).

Because of the modest sample size, regression diagnostics (Bollen & Jackman, 1985) were scrutinised to determine if the observed effects were being generated by one or a few influential cases. An influential case was so designated if it had a large studentised deleted residual, a large leverage value, and a large value of Cook's *D*. For the regression analysis of TC, no case had large values on all three diagnostics. For the regression analysis of HDL-C, there were two cases that appeared influential, primarily as a result of large leverage values. Leverage values indicate a case with a distinctive profile of values on the independent variables in the regression analysis; the two cases identified in the HDL-C analysis both had distinctive dietary patterns, but deletion of the cases made no difference in the results. Two influential cases were identified for the analysis of the TC : HDL-C ratio. One of these had an extremely large TC : HDL-C ratio (TC : HDL-C = 7.62). The other case was distinctive only by being at the extreme of the age distribution (age = 69). The deletion of these two cases did attenuate the interaction effect between lifestyle incongruity and kin support, but made no other difference in the results.

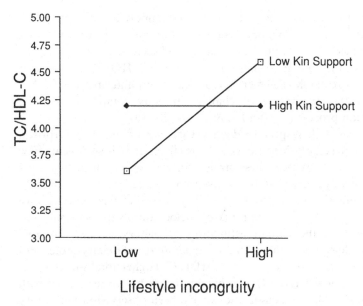

Fig. 2.3 TC : HDL-C ratio in relation to lifestyle incongruity by kin support.

Discussion

Before proceeding to a general discussion of these results, it is useful to consider both the methodological strengths and the weaknesses of this study. With respect to methodological weaknesses, sampling stands out. The sample size is very small, but more important than that, the true extent to which this sample represents the range of sociocultural and dietary variability in urban Brazil is difficult to assess. In ethnographic terms, the four groups sampled make a great deal of sense, but the question of the extent to which these groups may over-emphasise diversity, and hence over-estimate the effects of various factors, cannot be answered by using these data. This question requires replication as an answer.

Two major strengths stand out. First, sociocultural variables have been operationalised in accordance with a specific theory, one that takes into account both the variety of social and behavioural factors described as salient in the modernisation process, and also the complex interplay of external, macrosocial processes and traditional cultural processes (Dressler, 1990). Second, dietary intake variables have been measured so as to take into account the difficult problem of intra-individual dietary variability (Keys, 1988). By averaging across 24 h recalls, the accuracy and stability of estimated dietary intake is considerably enhanced.

With these caveats in mind, these results support a model in which physiological responses to modernisation and urbanisation are a function of the combined effects of physical, dietary, and sociocultural factors. The association of the body mass index and TC and TC:HDL-C is consistent with previous research (Pelletier & Hornick, 1986) and provides further evidence of the importance of changes in body composition in the modernisation process (see Bindon & Zansky, 1986).

The analysis with respect to dietary variables offered few surprises, unless the mere fact of observing such large effects at all is surprising. Few field studies have observed these straightforward effects of dietary variables, presumably because of the relatively constrained range of variability in dietary practices and the methodological difficulties noted above (Connor *et al.*, 1978). The wide range of sociocultural variability sampled here, along with the careful estimation of dietary intake, probably combined in detection of these effects. Higher intake of dietary cholesterol is associated both with TC and with HDL-C. Higher total fat intake is associated with lower HDL-C, and higher dietary fibre is associated with higher HDL-C. All these associations have been observed in laboratory and clinical studies (Grundy, 1986). It is worth noting that no dietary variable was associated with the ratio of TC to HDL-C.

The novel results of this study are the association of sociocultural factors with serum lipid values, independently of covariates and dietary factors. For TC, only higher socioeconomic rank was a significant correlate. The more elaborate model of lifestyle incongruity and the modification of its effects by kin and non-kin support were significant for HDL-C and TC:HDL-C, and both sets of results were as predicted. The process involved here clearly does not operate through behavioural changes associated with diet, since sociocultural variables and dietary variables are associated independently with lipids. Nor is the effect of sociocultural factors mediated by higher caloric intake and lower physical activity, since inclusion of the body mass index as a covariate makes no difference in the results. This suggests that the individual located in a social matrix described by high social incongruity and low social support responds in a psychophysiologically distinctive way, which in turn influences lipid values.

Elsewhere it has been argued that the individual with high incongruity and low support is repeatedly exposed to situations that lead to high levels of physiological arousal and cardiovascular reactivity (Dressler, 1992). The incongruous individual is presenting a sense of himself in social interaction as a high-status individual, based on his style of life. He anticipates an appropriate response from others, a response that is unlikely to be forthcoming, because others do not see him as being of high status on the basis of his low occupational class standing. In other words, he scans the social milieu for evidence that is not forthcoming. This kind of vigilant coping has been shown in laboratory work to be related to a variety of physiologic responses, which can in turn be compounded by the social isolation of low social support. Henry (1982) reviews the evidence that these social–environmental parameters are associated with changes in the sympathetic adrenal–medullary axis, the pituitary adrenal–cortex axis, and the hypothalmic–pituitary sex steroid axis.

These physiological processes may in turn influence how the intakes of dietary fats and cholesterol are metabolised, the result being varying relative levels of TC (or, most likely, LDL-C) and HDL-C. Given the pattern of results obtained here, it is tempting to speculate that given levels of total cholesterol are largely a function of dietary intake of cholesterol and fat, but that the specific subfractions of total cholesterol are a function of lipid metabolism determined in part by the social context of the individual.

This of course *is* speculation. The more important point to be emphasised is that there is a growing body of evidence linking specific social processes to biological parameters. An individual's physiological state, and hence his risk of chronic disease, are in part determined by his social adjustment to the urban environment and all of its demands and constraints. Future work must refine these observations and their meaning.

Acknowledgements

This research was suported in part by funds from PAHEF (the Pan American Health Education Foundation) and from The University of Alabama.

References

Allain, C. C., Poon, L. S., Chan, C. S. G., Richmond, W. & Fu, P. C. (1974). Enzymatic determination of total serum cholesterol. *Clinical Chemistry* **20**, 470–5.

Barnicot, N. A., Bennett, F. J. & Woodburn, J. C. (1972). Blood pressure and serum cholesterol in the Hadza of Tanzania. *Human Biology* **44**, 87–116.

Bindon, J. R. & Zansky, S. (1986). Growth and body composition. In *The Changing Samoans: Behaviour and Health in Transition* (ed. P. T. Baker, J. M. Hanna & T. S. Baker), pp. 222–53. New York: Oxford University Press.

Bollen, K. A. & Jackman, R. W. (1985). Regression diagnostics: an expository treatment of outliers and influential cases. *Social Methods and Research* **13**, 510–42.

Bucolo, G. & David, H. (1973). Quantitative determination of serum triglycerides by the use of enzymes. *Clinical Chemistry* **19**, 476–82.

Connor, W. E., Cerqueira, M. T., Connor, R. W., *et al.* (1978). The plasma lipids, lipoproteins, and diet of the Tarahumara Indians of Mexico. *American Journal of Clinical Nutrition* **31**, 1131–42.

Dressler, W. W. (1982). *Hypertension and Culture Change: Acculturation and Disease in the West Indies.* South Salem, New York: Redgrave Publishing Company.

Dressler, W. W. (1984). Social and cultural influences in cardiovascular disease: a review. *Transcultural Psychiatric Research Review* **21**, 5–42.

Dressler, W. W. (1990). Culture, stress, and disease. In *Medical Anthropology: A Handbook of Theory and Method* (ed. T. M. Johnson & C. F. Sargent), pp. 248–67. Westport, CT: Greenwood Press.

Dressler, W. W. (1992). Social and cultural dimensions of hypertension in blacks: underlying mechanisms. In *Pathophysiology of Hypertension in Blacks* (ed. J. G. Douglas & J. C. S. Fray), pp. 00–00. New York: Oxford University Press. (In press.)

Dressler, W. W., Dos Santos, J. E. & Viteri, F. E. (1986a). Blood pressure, ethnicity, and psychosocial resources. *Psychosomatic Medicine* **48**, 509–19.

Dressler, W. W., Dos Santos, J. E., Gallagher, P. N. Jr. & Viteri, F. E. (1987a). Arterial blood pressure and modernization in Brazil. *American Anthropologist* **89**, 389–409.

Dressler, W. W., Mata, A., Chavez, A. & Viteri, F. E. (1987b). Arterial blood pressure and individual modernization in a Mexican community. *Social Science and Medicine* **24**, 679–87.

Dressler, W. W., Mata, A., Chavez, A., Viteri, F. E. & Gallagher, P. N. (1986b). Social support and arterial blood presure in a central Mexican community. *Psychosomatic Medicine* **48**, 338–50.

Graves, T. D. & Graves, N. B. (1980). Kinship ties and the preferred adaptive strategies of urban migrants. In *The Versatility of Kinship* (ed. L. S. Cordell & S. J. Beckerman), pp. 195–217. New York: Academic Press.

Grundy, S. M. (1986). Cholesterol and coronary heart disease: a new era. *Journal of the American Medical Association* **256**, 2849–58.

Henry, J. P. (1982). The relation of social to biological processes in disease. *Social Science and Medicine* **16**, 369–80.

James, S. A., de Almeida-Filho, N. & Kaufman, J. S. (1991). Hypertension in Brazil: a review of the epidemiological evidence. *Ethnicity and Disease* **1**, 91–8.

Keys, A. (1988). Diet and blood cholesterol in population surveys. *American Journal of Clinical Nutrition* **48**, 1161–5.

Kostner, G. M. (1976). Enzymatic determination of cholesterol in high density lipoprotein fractions prepared for polyanion precipitation. *Clinical Chemistry* **22**, 695.

Labarthe, D., Reed, D. & Brody, J. et al. (1973). Health effects of modernization in Palau. *American Journal of Epidemiology* **98**, 161–74.

Page, L., Damon, A. & Moellering, R. C. (1974). Antecedents of cardiovascular disease in six Solomon Island societies. *Circulation* **49**, 1132–46.

Pelletier, D. L. & Hornick, C. A. (1986). Blood lipid studies. In *The Changing Samoans: Behavior and Health in Transition* (ed. P. T. Baker, J. M. Hanna, & T. S. Baker), pp. 327–49. New York: Oxford University Press.

Rosnow, R. L. & Rosenthal, R. (1989). Statistical procedures and the justification of knowledge in psychological science. *American Psychologist* **44**, 1276–84.

Straubhaar, J. D. (1989). Mass communication and the elites. In *Modern Brazil:*

Elites and Masses in Historical Perspective (ed. M. L. Conniff & F. D. McConn), pp. 225–45. Lincoln, Nebraska: University of Nebraska Press.

Thomas, P. D., Goodwin, J. M. & Goodwin, J. S. (1985). Effect of social support on stress-related changes in cholesterol level, uric acid level, and immune function in an elderly sample. *American Journal of Psychiatry* **142**, 735–7.

Weidner, G., Sexton, G., McLellareau, R., Connor, S. L. & Matarazzo, J. D. (1987). The role of Type A behavior and hostility in an elevation of plasma lipids in adult women and men. *Psychosomatic Medicine* **49**, 136–45.

3 The urban disadvantage in the developing world and the physical and mental growth of children

F. E. JOHNSTON

In *Stability and Change in Human Characteristics*, Bloom (1964) described the environment as 'the conditions, force, and external stimuli which ... surround, engulf, and play upon the individual' (p. 187). When there is a high probability that all members of a group will come under its influences, an environment may be said to be 'powerful' and the distributions of susceptible traits will shift in accordance with its pressures.

It is this powerful environment that comprises the urban disadvantage of the Third World urban poor. Given its potential for exerting major negative effects on child development, for increasing the risks to the young of morbidity and mortality, and for reducing their capacity as adults to generate capital, the first step in alleviating the effects of powerful environments is to determine how they affect physical growth and cognitive development. This presentation focuses on these issues using, as an example, an ongoing study of children of a socioeconomically marginal community on the periphery of Guatemala City.

The study and its setting

In 1978–9, a team of investigators from the Universidad del Valle de Guatemala and the University of Pennsylvania designed a study aimed at examining the effects of the environment on the physical growth, cognitive development, and academic achievement of the children of El Progreso, a community located on the outskirts of urban Guatemala City. With a population at the time of approximately 7500 inhabitants, El Progreso had been formed to help house those made homeless by the devastating earthquake of 1976. Data collection commenced in 1980 and has continued annually since then, focusing on three cohorts of children who, in the first year of the study, were 3, 5 and 7 years old respectively.

Johnston *et al.* (1987) and Johnston & Low (unpublished MS) describe the research design in detail; it will only be outlined here. The conceptual

26

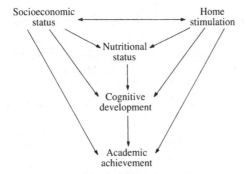

Fig. 3.1 Conceptual model of the study.

model underlying the study is presented in Fig. 3.1. Based on that model, seven categories of data have been collected annually:

1 physical growth;
2 cognitive development;
3 behavioural stimulation within the home;
4 academic achievement and school performance;
5 social, economic, and demographic characteristics of the home;
6 ethnographic and observational;
7 other data collected for specific purposes.

Four hundred and seventy children, selected randomly from census records, have been examined annually, within a few days of their birthdays.

Socioeconomic characteristics of El Progreso

El Progreso is a community not unlike many others of Guatemala and the rest of the developing world. With them, it exhibits the characteristics associated with Third World urban poverty. The distributions of selected variables are given in Table 3.1, and reveal low levels of education and high illiteracy. Only 3% of the mothers and 14% of the fathers had completed the six years of elementary school. As indicated, 49% of the homes had dirt floors. Most contained one or two rooms, and one third utilized kindling (i.e. *leña*) as a cooking fuel. The low level of material wellbeing is illustrated further in Fig. 3.2, which presents the distribution of the number of electrical appliances among the homes of the study sample. Although electrical service was available to each home, 43 (9%) had no electrical appliances and another 101 (21.5%) had but one (invariably a radio).

Contraceptive use is relatively low, as indicated by the mothers of cohort 5, from whom information was available; over one third reported never using contraception. The distribution of numbers of offspring ranged from

Table 3.1. *Selected indicators of SES in 470 families from El Progreso, study year 1*

	Fathers		Mothers	
	n	%	n	%
Migration				
Migrant	347	73.8	334	71.1
Non-migrant	123	26.2	136	28.9
Literacy (self-reported)				
Non-literate	45	9.6	142	30.3
Literate	425	90.4	328	69.7
Schooling				
No school	72	15.4	161	34.3
Some school	398	84.6	309	65.7

Contraceptive use by mothers (n = 167)	n	%
Not used	61	36.5
Sometimes used	106	63.5

Number of offspring of mothers	Mean	S.D.
	4.4	2.3

Type of floor in house	n	%
Earth	232	49.4
Cement	212	45.1
Other	26	5.5

1 to 12, with a mode and median of 4. However, it must be remembered that these figures are not representative of the entire community, since having a child was a prerequisite for a mother's participating in the study. None the less, the number of women who were never pregnant is quite low in the community. Infectious disease is widespread; our data indicate that children between 1 and 2 years of age average 6 bouts of diarrhoea per year.

The structure underlying the distribution of socioeconomic and demographic characteristics was investigated through principal component analysis (Johnston *et al.*, 1989). The purpose was to eliminate redundancy and permit the development of a composite indicator of socioeconomic status (SES). In the analysis, the first principal component represented indicators of relatively higher affluence and levels of material wellbeing, with four variables loading between 0.56 and 0.71: (1), the number of electrical appliances in the home; (2), the type of fuel used for cooking (*leña*,

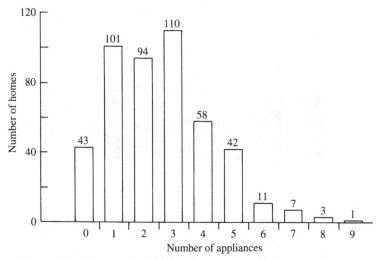

Fig. 3.2 Distribution of the number of electrical appliances in households, El Progreso, first study year.

bottled gas, or propane); (3), the number of years of schooling of the father; (4), the number of years of schooling of the mother. These variables were scaled into either quartiles or tertiles, as appropriate to their distributions. For each household, the rankings of the four variables were summed, yielding an SES score that ranged from 4 to 14. This has been used as the basic measure of SES, with individual variables also employed to test specific hypotheses.

In the analysis, two other components had eigenvalues greater than 1.0. The second principal component resented parental birthplace, scaled in terms of distance from Guatemala City, and the third the type of house. However, these have not generally been utilised in the analysis.

Effects of socioeconomic status on growth, development and achievement

Physical growth

All physical growth data for the first five years of data collection were pooled, and separate multivariate analyses of variance (MANOVA) were carried out at 3, 5, 7, 9, and 11 years of age utilizing the following model:

(Height, Weight, BMI, HeadC, Triceps,

UAMC) = Constant + Cohort + Sex + SES,

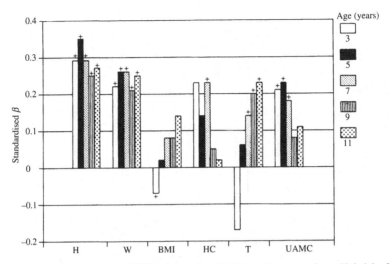

Fig. 3.3 Analysis of SES and growth at 3,5,7,9 and 11 years of age. H, height; W, weight; HC, head circumference; T, triceps skinfold; +, significant ($p < 0.02$).

where BMI = body mass index and UAMC = upper arm muscle circumference. Sample sizes ranged from 110 at age 3 to 424 at age 11. Fig. 3.3 presents the analysis of the effects of SES, expressed as standardised partial regression coefficients, on growth.

From 3 to 11 years of age, there is a clear effect of SES on growth in height and weight. Children from families with higher SES scores are taller and heavier. On the other hand, there is no relationship between SES and BMI. The depression in growth associated with a poor environment is relatively the same for both height and weight, leading to nutritional dwarfing rather than to a disproportionate wasting of tissue. While there are effects of SES on the composition of the arm, they are not as marked as for general body size. In the earlier years of the age range, UAMC shows a signifiant association; in the later years, the triceps skinfold shows the effect. There was essentially no association of SES with head circumference.

The indicator of SES employed here is a composite one; it might be argued that more subtle sociocultural distinctions are missed by combining measures of material possessions in the home with educational levels of the parents. While there is a correlation between schooling and material possessions, education may indicate other parental attributes that could affect the home environments of their children and affect growth and development. For this reason, the MANOVA was repeated, substituting the quartile or tertile rankings of the four components of SES for the

Table 3.2. *Significant* F *ratios from* MANOVAs *of body measurements on type of cooking fuel, adjusted for cohort and sex, by age group*

	Age Group (yr)				
	3	5	7	9	11
Variable					
Height	ns	*	***	**	**
Weight	ns	***	***	***	**
BMI	ns	**	*	**	ns
Head C	ns	ns	**	ns	ns
Triceps	ns	ns	**	**	**
UAMC	ns	**	**	*	ns
Multivariate					
F (Wilks)	ns	**	**	**	*
n	96	165	424	320	193

Asterisks indicate level of significance: *, $0.01 < p < 0.05$; **, $0.001 < p < 0.01$; ***, $p < 0.001$; ns, not significant.

composite indicator. The results indicated clearly that there was no independent relationship of growth and mother's educational level. Although the father's level of education was significant when it was alone in the model, it disappeared when either the type of cooking fuel or the number of electrical appliances was entered.

Table 3.2 shows the results of analyses of body measurements, again adjusted for cohort and sex. Since there was essentially no relationship to the educational level of either parent, the table presents only the significant F ratios for the type of cooking fuel. The patterns are, in general, similar to those obtained with the composite measure, though the results tend to be stronger. Using the composite measure, 18 to 30 beta coefficients are significant; using only the type of cooking fuel, 22 are signifcant.

The results suggest that socioeconomic status is associated significantly with growth during the middle years of childhood. The relationship is especially strong with height and weight, measures of general growth. The socioeconomic effects are mediated through measures of material well-being in households. While father's educational level is related to growth when it is the only SES variable in the model, the relationship disappears when indicators of material possessions are included. Father's education therefore acts as a proxy for the household rather than as an indicator of other attributes. There is no evidence of any relationship between a mother's characteristics and the growth of her children, except as those characteristics are predictive of the father or of the household.

Cognitive development

Cognitive development was assessed through the Wechsler IQ test, administered during the first five years by a single trained assistant in a building especially constructed in the community for examinations. The analyses reported here are based on full scale WISC IQs for every age except 3 years, when a modified version of WPPSI, the pre-school version, was employed. Since no standardisation was available, analysis at age 3 has used the raw score as the dependent variable.

As an example of the results, Fig. 3.4 presents the analysis of variance of IQ at age 7, with 5 independent variables: sex, composite SES, home stimulation score, mother's verbal IQ (WAIS), and child's height. The sample size is reduced because IQ scores were not available for all mothers. The probability levels, at the tops of the bars, reveal significant effects of SES, height, and stimulation score. Mother's IQ was not significantly related after adjusting for the other variables, and there was no sex difference.

The relationship between mother's IQ and the IQ of her child is clarified further by Table 3.3, which presents the results of the analysis of variance of IQ at age 7 by several maternal characteristics. Of importance is the significant interaction between the mother's tertile of educational attainment and her IQ, in determining the cognitive development of her child.

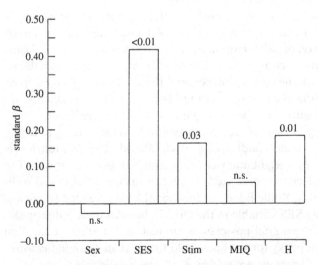

Fig. 3.4 Analysis of variance of IQ at 7 years of age. Stim, home stimulation; MIQ, mother's IQ; H, height; $n = 137$, $r^2 = 0.346$. Figures indicate significance levels; n.s., not significant.

Table 3.3. *ANOVA of maternal characteristics and child's IQ at age 7 (*n = *137,* r^2 = *0.246)*

Characteristics	F	p
IQ	1.146	ns
Education	0.487	ns
Literacy status	0.972	ns
Stimulation score	8.775	0.004
IQ/education	4.938	0.009

Table 3.4. *Regression of child's IQ at age 7 on mother's IQ by mother's educational level*

Educational level	n	Constant	b	F	p
1	42	−0.16	0.09	0.202	ns
2	51	−0.08	−0.11	0.431	ns
3	44	0.16	0.49	17.916	<0.001

Table 3.5. *ANOVA of WISC IQ on the components of SES in 7-year-olds, adjusted for sex (*n = *353,* r^2 = *0.206)*

SES Component	F	p
Mother's education	2.670	ns
Father's education	4.314	0.005
Number of electrical appliances	9.929	<0.001
Type of cooking fuel	0.582	ns

Because of the interaction, the IQ of the child at 7 was then regressed on that of its mother by tertile of mother's education; the results are shown in Table 3.4. There is a significant association only for those mothers with the highest levels of educational attainment, in El Progreso those who attended school for four or more years. Among mothers with no or only a few years of school, their IQ's were unrelated to those of their children.

In order to assess the relative influence of the components of SES on cognitive development, we have analysed them at age 7 by an analysis of variance. Table 3.5 shows the results. Only the educational level of the father and the number of electrical appliances, a measure of material wellbeing, are significant. Fig. 3.5 presents the mean IQ by quartile of these variables. The differences are quite striking, amounting to approximately ten points from the lowest to the highest quartile.

The final analysis of cognitive development presented here is given in Fig. 3.6, which shows the mean IQ at age 7 by home stimulation scores. The IQs are adjusted for cohort, sex, SES, stature, and head circumference and reveal a clear relationship independent of socioeconomic status. From the lowest to the highest levels of stimulation, the difference is almost 15 IQ points.

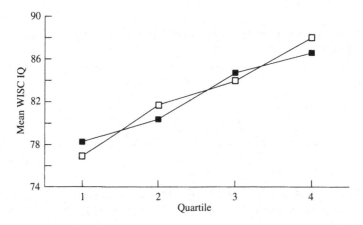

Fig. 3.5 Mean IQ at age 7, sex adjusted, by quartile of father's education (filled squares) and number of electrical appliances (open squares).

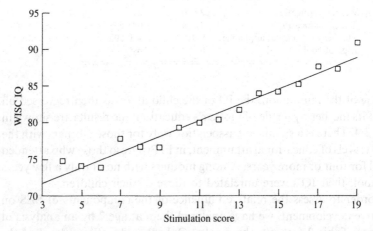

Fig. 3.6 Mean WISC IQ at age 7 by home stimulation score, adjusted for cohort, sex, SES, stature and head circumference.

The analysis of the relationships between the environments of the children of El Progreso and their cognitive development reveals two clusters of determinants. The first is socioeconomic, reflecting the material wellbeing of the home and correlated with the educational level of the father. The second cluster is behavioural, reflecting the interaction of mother and child, and independent of the material items in the home.

The relationship of the mother's IQ to that of her child is a complex one. As part of the network of socioeconomic characteristics of the families, a mother's IQ is but one indicator of economic status. However, the influence of a mother's IQ cannot be evaluated apart from her educational level. Among the mothers from the lower two thirds of years of schooling (less than four years), their IQs and those of their children are unrelated. However, for those mothers with four or more years of schooling, the relationship is positive, linear, and strong, with a correlation of 0.547. The factors underlying this interaction of IQ and education among the mothers cannot be explained completely. One interpretation is that, among the better-educated mothers of the sample, other factors are able to express themselves in ways not possible in mothers with little or no schooling. A second interpretation is that these factors may be attitudinal ones, not picked up in our data. Or, finally, they may be genetic, in that the genetic relationship between maternal and child IQ is not expressed in this community among the very poor and the uneducated.

Academic achievement

The last area to be discussed is academic achievement. Two categories of data are available for analysis. The first is reading achievement scores from standardised tests given in school at the end of the academic year. The second is the end-of-year grades in language and mathematics assigned by the teachers. Fig. 3.7 presents the results of the analysis of variance of reading achievement scores in grades 1, 2, and 3 by measures of growth, SES, stimulation, and cognitive development. The only variable that is significant in each grade is the WISC IQ. Though not presented here, the results are the same for the teacher-assigned marks in mathematics and reading.

These results indicate that only IQ exerts a significant direct effect upon scholastic performance of the children of El Progreso. However, this does not negate the role of the environment in academic performance since analyses already presented have demonstrated that both SES and stimulation in the home are significant determinants of IQ. The best conceptualisation of the environments of these children is an ecological one, in which there is a network of interactions consisting of direct, indirect, and joint

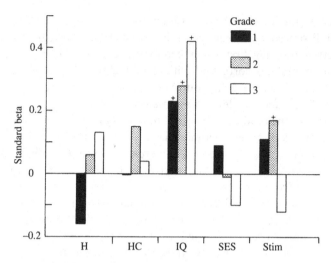

Fig. 3.7 Analysis of academic achievement and environmental factors, by grade in school. Values of r^2 for grades 1, 2 and 3 are 0.101, 0.206 and 0.269, respectively; +, significant ($p < 0.05$). Abbreviations as for Figs 3.3 and 3.4.

effects. All of these effects add up to a strong relationship between environmental variables and school performance.

Conclusion

Two clusters of environmental determinants are seen in El Progreso. The first one is statistically the most important and reflects the poverty and economic disadvantage which is ubiquitous there as well as throughout the urban Third World. Appearing as significant in virtually every analysis reported here, this component is represented by measures of the material possessions of the household, along with the educational level of the father. Other variables included in the study also fell into this component, such as those that described the type of house, but they were not as robust in accounting for growth and developmental outcomes. In any event, all of these variables are best conceptualised as indicators of the relative poverty (or affluence) within which children develop and further as indicators of nutritional and health status and those socioeconomic qualities of the environment that affect cognitive development.

The second cluster of environmental determinants is a maternal one, significant in the cognitive development of the children and seen primarily as the style of interaction between mother and child. In terms of its effects on development, this cluster is independent of the economic factors discussed above. Though not reported here, we have also shown these clusters to be important in the fertility of mothers and the survival of their children (Johnston *et al.*, 1989).

The data reported here suggest the primacy of socioeconomic variables as determinants of child development in the urban Third World. Within the context of Bloom's argument, social and economic status fulfil the criteria for an extreme powerful environment, and these analyses indicate the difficulty of escaping its effects. This finding, if generalisable to other settings, is important in demonstrating that programmes designed to increase nutritional and cognitive levels must first come to grips with the poverty and disadvantage that is ubiquitous in such settings. While other determinants, such as mother–child interaction, are also important, the findings of this study show clearly that the urban disadvantage is a material one, whose effects extend through growth and cognition to academic achievement and, by extension, the ability to generate capital as an adult.

References

Beaujier-Garnier, J. (1978). Large overpopulated cities in the underdeveloped world. In *Geography and a Crowding World* (ed. W. Zelinsky, L. A. Kosinski & R. M. Prothero), pp. 269–78. New York: Oxford University Press.

Bloom, B. S. (1964). *Stability and Change in Human Characteristics*. New York: Wiley.

Colburn, F. D. (1986). De Guatemala a Guatepeor. *Latin American Research Review* **21**, 242–8.

Johnston, F. E. & Low, S. M. *The Biosocial Ecology of Physical Growth, Cognitive Development, and School Achievement in a Disadvantaged, Urban Guatemalan Community*. (Unpublished manuscript.)

Johnston, F. E., Low, S. M., deBaessa, Y. & MacVean, R. B. (1987). Interaction of nutritional and socioeconomic status as determinants of cognitive development in disadvantaged, urban Guatemalan children. *American Journal of Physical Anthropology* **73**, 501–6.

Johnston, F. E., Reid, W., deBaessa, Y. & MacVean, R. B. (1989). Socioeconomic correlates of fertility, mortality, and child survival in mothers from a disadvantaged, urban Guatemalan community. *American Journal of Human Biology* **1**, 25–30.

4 Differences in endocrine status associated with urban–rural patterns of growth and maturation in Bundi (Gende-speaking) adolescents of Papua New Guinea

B. ZEMEL, C. WORTHMAN AND C. JENKINS

Urban residence is frequently associated with increased growth and maturation. However, the mechanisms by which the urban environment affects growth and development are still not well understood. In this study, the role of steroid hormones in explaining urban–rural differences in adolescent growth is explored. One hundred and twenty-nine urban and 409 rural Bundi adolescents (aged between 10 and 24 years) were examined cross-sectionally in 1983–4. Growth status was assessed and serum samples were analysed for testosterone (T), dehydroepiandrosterone (DHEA) and its sulphate (DHEAS), androstenedione (A), oestrone (E1), oestradiol (E2), and progesterone. Urban youths are significantly ($p \leq 0.05$) taller and heavier, with greater body mass index than rural youths. These differences are greatest at younger ages and less or not significant at older ages. Comparable steroid secretory patterns were observed. Urban girls, 11–15 years old, have significantly greater steroid levels for all measures. For males, significant urban–rural differences are observed in the 16–20 year age range for D, DS, E1, and A. At older ages, there are no urban–rural differences. Using stepwise regression analysis, endocrine measures (E1, A, P) explained 54% of the variation in height for females. For males, T, E1 and/or E2, and A explained 65–76% of the variation in the growth status and urban residence explained a small (1%), but statistically significant, amount of additional variation. These findings suggest that earlier maturation is the most prominent feature of the growth of urban Bundi adolescents.

Introduction

Many studies of the effects of urbanisation on childhood growth and development have shown that urban residence is associated with increased

38

growth status (Zhang & Huang, 1988; Bielicki & Welon, 1982; Bogin & MacVean, 1981; Malina, 1979; Malcolm, 1975). Age at menarche and the onset of the adolescent growth spurt also occur at earlier ages in urban environments (Bogin, 1988; Laska-Mierzejewska *et al.*, 1982). Ecologically, this proclivity for improved growth and maturation in urban environments has several important implications. First, when an urban environment is not conducive to adequate or improved growth, adverse living conditions such as severe overcrowding and poverty outweigh any positive benefits of urban life (Johnston *et al.*, 1985; Malina *et al.*, 1981). In this regard, the growth of urban children is a useful indicator of the health and wellbeing of urban populations and the quality of the urban environment (Bielicki, 1986). Secondly, where urbanisation is a recent and rapidly occurring phenomenon, the sociocultural and economic mechanisms for incorporating maturing adolescents into the roles of adults may not be able to keep pace with these patterns of biological change at the population level (Worthman & Whiting, 1987; Worthman, 1987). 'Problem behaviours', such as involvement in gangs and teen pregnancy, exemplify difficult outomes of this discordance. Similarly, the increase in reproductive life span due to earlier sexual maturation can result in rapid population growth. Thus, growth and maturation patterns should be an important consideration in assessing the needs, health status and demography of urban populations.

The mechanisms by which the urban environment affects growth and development are still not well understood. Easier access to health care, improved sanitation, nutrition, educational and recreational facilities, and decreased energy expenditure are widely considered causal factors; however, the underlying biological processes remain unclear. In this study, the role of steroid hormones in explaining urban–rural differences in adolescent growth is explored. In late childhood and adolescence, rapid changes in adrenal and gonadal steroid hormones levels are associated with sexual and skeletal maturation, the adolescent growth spurt, the appearance of secondary sexual characteristics, and accompanying behavioural changes. The adolescent secretory patterns of these hormones have been well documented for Western children (Nottelman *et al.*, 1987; Preece *et al.*, 1984; Apter, 1980; Lee *et al.*, 1976; Lee & Migeon, 1975; Sizonenko *et al.*, 1976; Sizonenko & Paunier, 1975), but little is known about their variation in differing populations and environments worldwide.

Both adrenal androgens and gonadal steroids play a role in pubertal development. Gonadal steroid hormone concentrations reflect, of course, the degree of activation of the gonad in the course of puberty, but these hormones also influence epiphyseal maturation and thereby shape advancing skeletal maturation (Wierman *et al.*, 1986). For males, testosterone is produced primarily by the testes and is thus a good indicator of gonadal maturation. In females, oestrone and oestradiol are mainly products of the ovaries, progesterone is secreted by the corpus luteum, and the primary source of testosterone is the adrenal gland. Although ovarian secretory patterns become cyclical as puberty progresses, they are useful indicators of gonadal functioning at the population level. The principal androgen secreted by the adrenal in both males and females is DHEAS, which is used here as an indicator of adrenal androgen production. Other important adrenal androgens are DHEA and androstenedione. In women, adrenal androgens are produced by both the adrenals and the ovaries (Kirschner *et al.*, 1973); the latter may be the main source of testosterone and androstenedione in puberty (Apter *et al.*, 1982). Production of each of these adrenal hormones increases rapidly and fairly linearly during puberty; however, their functional roles are not well understood. Because of the emerging cyclicity of ovarian steroids, adrenal androgens are the best linear correlates of chronological age and pubertal progression in females.

In males, and possibly in females, testosterone is the most potent androgen and is responsible for much of the increase in height, muscle mass and skeletal maturation during adolescence (Zemel, 1989; Bing *et al.*, 1988; Boepple *et al.*, 1988; Nottelman *et al.*, 1987; Zemel & Katz, 1986; Raisz & Kream, 1981; Winter, 1978; Tanner *et al.*, 1976; Blizzard *et al.*, 1974). Estrogens, in small doses, also are associated with linear growth (Laue *et al.*, 1989; Zachmann *et al.*, 1986; Caruso-Nicoletti *et al.*, 1985). DHEAS has been shown to explain additional variation in growth velocity and height status, after accounting for the effects of testosterone (Zemel, 1989; Zemel & Katz, 1986), suggesting that both adrenal and gonadal steroids contribute to growth in height during normal puberty.

This study of adolescent growth and endocrine development is part of a larger survey of health and nutrition among the Bundi of Papua New Guinea (PNG). Previous studies of the Bundi in 1966–7 showed a pattern of extreme growth delay, small body size and retarded sexual maturation (Malcolm, 1970). This study represents the first of its kind in describing endocrine variation in a population with delayed growth, and in exploring the effects of the urban environment on growth and endocrine development in a non-Western population.

Sample and methods

A cross-sectional sample of adolescents was interviewed and examined in 1983–4 in the cities of Goroka, in the Highlands, and Madang, on the coast, in addition to rural youths in schools and villages. In general, urban Bundi live in settlements located in undeveloped urban areas. Although more densely populated, urban settlements are reminiscent of rural villages.

Houses are constructed of bush materials, wood and/or tin, with earthen floors, firepits for cooking, and little or no furniture. Other urbanites live in Western-style houses with electricity and indoor plumbing. All urban settlements differ from rural villages in higher population density and absence of land for extensive gardens.

The urban Bundi are varied in migration history, educational and economic achievement, and location. Circular migration between villages and cities is common, but even long-term city-dwellers, some having been in residence for forty years, maintain strong social and economic ties with their villages (Zimmer, in press). The urban sample includes students at a boarding school. The school is not located in a city, but many of the students had previously lived in cities; in most respects, life at the school resembled urban living.

Although urban lifestyles are varied, certain common features of urban life are distinct from rural life. In urban areas, the proximity of goods and service, and the existence of paved roads and public transportation, contribute to reducing energy expenditure, in addition to more sedentary occupations. Health care facilities and the availability of medicines greatly improve health status. Urbanites are almost entirely dependent on store-bought foods dense in calories, fat and protein, and lower in fibre. The urban staples are rice, tinned fish or meat, biscuits, and tea with sugar, as opposed to the traditional rural diet of sweet potatoes and leafy greens. In contrast, rural Bundi is an area of steep mountains and rough terrain. Rural villages range in altitude from 1030 to 2100 m, which is below the altitude at which hypoxic effects on growth are observed (Harrison *et al.*, 1988). Most villages are not accessible by car or truck, and electricity and running water generally are not available. Gardening is the major activity; villagers walk long distances, often carrying heavy loads from their gardens, as part of their daily routine.

The samples consist of 129 urban and 409 rural adolescents between the ages of 10 and 24 years. Of these, birthdates were known for 63 urban and 282 rural adolescents. At the time of measurement, the age of the subject was not known since chronological age is not commonly reckoned by the Bundi. The baptismal records at the Bundi Catholic Mission were the primary and most reliable source of birthdates. For the urban sample, birthdates were more difficult to obtain; only 50–55% of these subjects had known birthdates. Chronological age is used in some, but not all, of the analyses that follow. From this larger sample, endocrine data are available for 80 urban and 212 rural adolescents, with known birthdates for 42 urban and 151 rural adolescents.

The protocol began with a personal interview, followed by an anthropo-metric examination, including height, weight, triceps and subscapular skinfold thickness, and biacromial and bi-iliac diameters. Blood was drawn for assessing steroid hormone levels and analysed by radioimmunoassay according to standard techniques (Wen *et al.*, 1985; McNatty *et al.*, 1979*a*, *b*; Manlimos & Abraham, 1975; Abraham *et al.*, 1972). Results are presented for testosterone, dehydroepiandrosterone (DHEA) and its sulphate (DHEAS), oestrone, oestradiol, androstenedione, and progesterone.

All statistical analyses that follow were conducted using the Statistical Analysis System (SAS). For the two-way analysis of variance models, the General Linear Models (GLM) procedure was used to account for unbalanced design.

Results

Urban–rural differences in growth and age at menarche

Although there has been a significant secular trend in growth in the last twenty years, rural Bundi children and adolescents, on average, are still below the NCHS fifth percentile for height at all ages, and for weight at most ages, as are most children in PNG (Jenkins *et al.*, 1987, 1990; Zemel & Jenkins, 1989). Sexual maturation and the completion of physical growth are delayed. Consequently, for the purposes of this study, urban Bundi are compared with rural Bundi children to provide more meaningful comparisons in the evaluation of any environmental effects. A two-way analysis of variance model was used to test the effects of urban versus rural residence and age group on measures of growth status. An interaction term was included to test the hypothesis that the differences between urban and rural groups are not the same across all ages. Because of the relatively small sample size of the urban group with known ages, initial urban–rural comparisons are made on the basis of the age groupings 6–10, 11–15, 16–20 and 21–24 years. Statistically significant differences ($p \leq 0.05$) are observed according to age group and residence (Fig. 4.1), such that for both males and females, urban youths are taller and heavier, with greater body mass index than rural youths. The age group–residence interaction term was also statistically significant, indicating that the urban–rural differences in growth are not parallel; the differences are greater at the younger ages and lesser or not significant at older ages.

For males 6–10 years old, there is a statistically significant difference in age between the urban and rural groups. The rural boys are more than a year older on average (6.9 versus 8.2 years), so it is difficult to make interpretations about this youngest age group. For the older age groups, there are clear and highly significant urban–rural differences in growth for

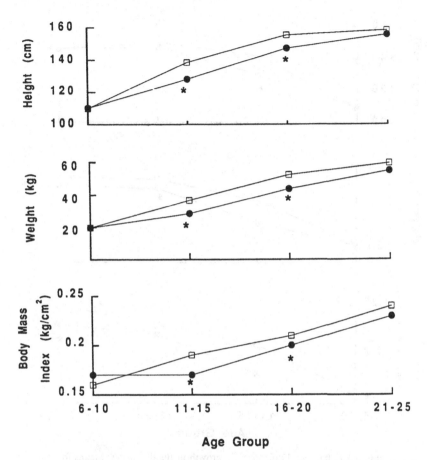

Fig. 4.1 Urban–rural differences in growth of Bundi males. Means for height, weight and body mass index by age group showing significant urban–rural differences at ages 11–15 and 16–20 years. Note that the 6–10-year-old groups differ in mean age, so urban–rural differences in this age range could not be determined. Symbols: circles, rural; squares, urban; *, significant difference in adjusted means ($p \leq 0.05$).

the 11–15 and 16–20 year age groups ($p \leq 0.05$). Urban boys 11–15 years old are on average 10 cm taller and 8 kg heavier than rural boys. Among the 21–24 year olds, there is a tendency for urban males to be taller and heavier, but the only statistically significant urban–rural difference is in triceps skinfold thickness.

For females, the urban–rural differences are age-dependent (Fig. 4.2). In the youngest age group there are no differences according to residence. Among 11–15 year olds there are significant differences in almost all measures of body size. The urban girls are significantly taller by over 9 cm

44   B. Zemel et al.

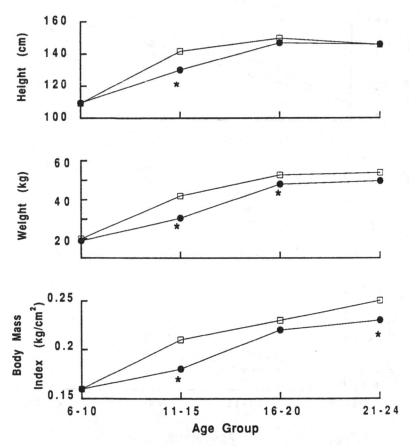

Fig. 4.2 Urban–rural differences in growth in Bundi females. Means for height, weight and body mass index by age group showing significant urban–rural differences in all measures at ages 11–15 years, with significant differences in weight and body mass index for older age groups. (Symbols: as Fig. 4.1.)

and heavier by 9 kg. They also have significantly greater body mass index. Among 16–20 year olds, the urban girls are only slightly taller, but they are significantly heavier. With the exception of body mass index, there are no urban–rural differences among the 21–24 year olds.

For both males and females, the magnitude of the urban–rural differences in growth status in the mid-adolescent age range suggests differences in the timing of sexual maturation and the adolescent growth spurt. Estimates of age at menarche for urban and rural girls support this hypothesis. Using a probit analysis of status quo data on menarche, the median age at menarche for rural girls is estimated to be 17.23 years (16.53 to 18.05 years, $n = 107$). Malcolm (1970) estimated the median age at menarche to be 18.0 years in

1966–7. Although the median age of menarche has decreased, rural Bundi girls still have one of the latest menarcheal ages ever observed. For urban girls, age at menarche is 15.78 (13.87 to 18.11 years, $n = 25$). Because of the small number of urban girls with known birthdates, it is difficult to obtain a precise estimate of the median age at menarche; thus the confidence ranges of the two estimates overlap. However, this difference of one and a half years in the estimated median age at menarche between urban and rural girls suggests that the urban–rural differences in adolescent growth, especially in the 11–15 year age range, can be explained by an earlier growth spurt and earlier sexual maturation among urban girls. This is also supported by the endocrine data presented below.

Urban–rural differences in endocrine status

Using the same design as above, a two-way analysis of variance was used to determine the effect of residence and age on steroid hormone levels. Log-normalised values of the hormone were used in the analysis.

For females, the overall models were all highly significant ($p \leq 0.01$), except for progesterone. For DHEAS and DHEA, residence was not a statistically significant effect, although the interaction of age and residence was significant or marginally significant at $p \leq 0.05$ and $p = 0.06$, respectively. Comparisons of the adjusted means show significant ($p \leq 0.05$) differences in DHEAS and DHEA for the 11–15 year old age group. For oestrone, the age–residence interaction term was significant along with the main effect. Fig. 4.3 shows the log values used in the analysis. The mean steroid levels are given in Table 4.1. Among rural girls there is a significant increase in mean concentrations between the 11–15 year old and the 16–20 year old age groups for all steroids except DHEAS and progesterone. For urban girls, only androstenedione exhibits a significant age increase at these same ages. Urban girls have greater mean steroid levels for all measures. These differences are statistically significant (except progesterone) for the 11–15 year old age group, but not at the older ages. Although these data are cross-sectional, they indicate that 11–15 year old urban girls are similar endocrinologically to older urban girls, whereas among rural girls, significant changes are occurring across these ages. Combined with the growth and menarche data, these findings suggest that earlier maturation is a major component of urban–rural differences in Bundi females.

For males, the number of urban 11–15 year olds with known ages is too small to include in the analysis of variance, but their mean log values are included in Fig. 4.4. The mean hormone concentrations are given in Table 4.1. With the exception of progesterone, all the steroids show significant increases in mean levels between ages 16–20 and 21–24 in the rural group,

Table 4.1. *Steroid hormone concentrations for urban and rural Bundi by age group*

Males	11–15 years				16–20 years				21–24 years			
	rural (n=47)		urban (n=3)		rural (n=31)		urban (n=15)		rural (n=12)		urban (n=5)	
	mean	s.e.	mean	s.e.	mean	s.e.	mean	s.e.	mean	s.e.	mean	s.e.
Testosterone[a]	264.3	59.8	9680.5	7356.5	3499.7	605.1	5693.3	1218.8	9501.1	2183.8	6863.0	978.9
DHEAS[b]	385.2	81.8	1079.3	513.1	791.3	89.2	1202.0	134.7	1452.7	252.2	1800.2	412.3
DHEA[a]	520.5	72.9	1350.3	408.9	1001.3	147.2	1418.3	168.9	1843.9	488.2	2210.4	573.7
Oestrone[a]	8.5	0.5	45.0	18.2	13.9	1.6	32.6	5.1	36.6	5.5	40.8	6.9
Oestradiol[a]	17.1	0.7	29.7	11.0	26.2	1.5	31.5	3.4	46.2	3.5	46.0	2.9
Androstenedione[a]	162.6	15.8	545.7	327.4	375.5	38.9	611.8	50.2	635.0	47.9	1214.2	199.9
Progesterone[a]	98.5	10.3	245.5	53.5	123.3	11.6	148.9	22.5	144.2	23.0	171.2	23.4
Females	rural (n=28)		urban (n=8)		rural (n=25)		urban (n=6)		rural (n=4)		urban (n=4)	
	mean	s.e.	mean	s.e.	mean	s.e.	mean	s.e.	mean	s.e.	mean	s.e.
Testosterone[a]	206.0	23.1	256.6	33.9	414.3	39.6	433.3	35.9	350.0	57.6	362.5	80.4
DHEAS[b]	351.8	58.4	988.3	144.8	1181.1	141.1	1180.8	107.8	1389.0	556.4	924.5	95.2
DHEA[a]	984.6	190.2	2049.4	401.3	2591.8	225.3	2220.0	341.9	3228.3	589.9	3160.3	774.9
Oestrone[a]	14.0	1.3	50.4	5.6	41.3	5.1	58.8	11.4	37.5	8.1	43.0	6.8
Oestradiol[a]	28.8	2.9	71.9	11.0	84.2	11.0	117.5	34.3	59.0	11.4	76.2	23.7
Androstenedione[a]	363.1	42.7	583.5	80.2	918.1	87.3	1309.5	149.5	854.8	253.0	1217.0	249.3
Progesterone[a]	132.9	15.7	260.6	71.3	268.0	66.6	678.7	359.8	674.3	565.4	229.0	20.6

Notes:
[a] (pg ml^{-1})
[b] (ng ml^{-1})

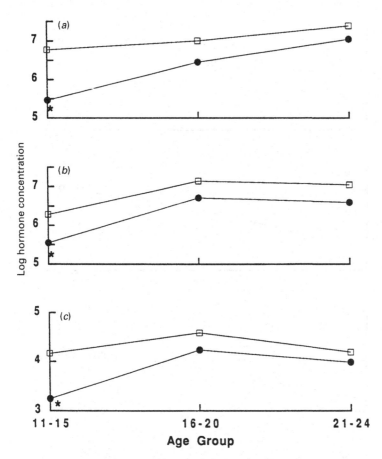

Fig. 4.3 Urban–rural differences in steroid hormones by age group for Bundi females. Means for the natural logarithms of concentrations of (a) DHEAS (ng ml^{-1}), (b) androstenedione (pg ml^{-1}) and (c) oestradiol (pg ml^{-1}) by age group showing significant urban–rural differences at ages 11–15 years. (Symbols: as Fig. 4.1.)

but not in the urban sample. On the average, urban males have greater mean androgen levels for all age groups. Statistically significant urban–rural differences occur in the 16–20 year old age range for DHEAS, DHEA, oestrone, and androstenedione. There are no significant urban–rural differences in testosterone for males, although at ages 16–20 the differences are marginally significant ($0.05 \leq p \leq 0.10$). In all, this pattern of differences in endocrine status suggests that urban males mature earlier; by the ages of 16–20 years, they are not significantly different from young adults, whereas the rural males are still undergoing rapid changes at these ages.

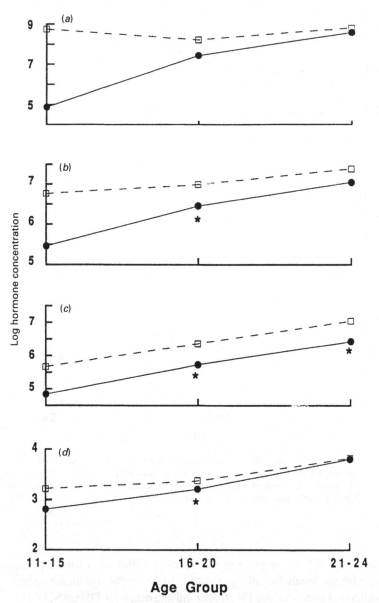

Fig. 4.4 Urban–rural differences in steroid hormones by age group for Bundi males. Means for the natural logarithms of concentrations of (*a*) testosterone (pg ml^{-1}), (*b*) DHEAS (ng ml^{-1}), (*c*) androstenedione (pg ml^{-1}) and (*d*) oestradiol (pg ml^{-1}) by age group showing significant urban–rural differences at ages 16–20 years. The number of 11–15-year-old urban males was too small to include in the analyses, but their mean values are shown here. The urban–rural differences in testosterone for the 16–20 year age group are marginally significant ($0.05 \leq p \leq 0.10$). (Symbols: as Fig. 4.1.)

The increased growth status of urban Bundi adolescents is associated with earlier maturation as indicated by steroid secretory patterns and menarcheal age. The question remains as to whether the urban environment is associated with increased size, if one takes into account their earlier maturation. Because of the cyclical nature of sex steroid secretion in females, this analysis is not possible with the present data. However, for males, this can be addressed by using testosterone as an indicator of maturation. All urban and rural boys, regardless of known age, were grouped into categories of pre-puberty (testosterone < 1000 pg ml^{-1}), mid-puberty ($1000 \leq$ testosterone ≤ 3000) and post-puberty (testosterone > 3000). Again, the two-way analysis of variance was used to test the hypothesis that urban residence and maturation status each have an effect on steroid concentrations and growth. The results are illustrated in Fig. 4.5, with corresponding mean values in Table 4.2.

For the pre-pubertal groups, there are significant urban–rural differences in testosterone concentrations, although they do not differ in estimated age. Differences in testosterone concentrations are not apparent in mid-puberty and post-puberty. However, urban boys have greater mean concentrations of all the other hormones except oestradiol. DHEAS, DHEA, androstenedione, oestrone and progesterone are significantly greater among pre-pubertal urban boys. In mid-puberty, there are marginal urban–rural differences in DHEAS ($p = 0.07$), and significant differences ($p \leq 0.05$) for oestrone and oestradiol. In urban post-pubertal boys, androstenedione, DHEA, and oestrone are significantly greater than in rural boys. Similarly, urban–rural differences in growth are apparent for height, weight, and body mass index (Fig. 4.6), and are statistically significant for the pre-puberty and post-puberty groups.

These results indicate that not only do urban boys mature earlier, but they are also larger for the same stage of maturation. Furthermore, when boys are compared on the basis of testosterone level, the urban–rural differences in oestrogens and androgens persist. These hormones may contribute to explaining environmental variation in growth and maturation even though, biologically, their effects are believed to be weak in comparison to testosterone. Finally, a stepwise regression analysis was used to ascertain which steroids best explain variation in growth status, and whether urban versus rural environment has additional explanatory value once these biological relationships are accounted for. All study participants who were believed to be within the age range from 10 to 24 years were included in the analysis, even if their exact age was not known. Urban residence and all seven steroids were allowed to enter the models to predict height, weight and body mass index, for males and females separately.

Table 4.2. *Steroid hormone concentrations for urban and rural males by puberty status*

| | Pre-puberty | | | | Mid-puberty | | | | Post-puberty | | | |
	rural (n=72)		urban (n=6)		rural (n=18)		urban (n=5)		rural (n=37)		urban (n=32)	
	mean	s.e.	mean	s.e.	mean	s.e.	mean	s.e.	mean	s.e.	mean	s.e.
Testosterone[a]	195.3	27.6	386.3	109.0	1787.7	137.3	1964.6	365.3	8563.7	843.1	7394.7	630.2
DHEAS[b]	396.5	62.2	736.2	209.3	608.1	97.1	1236.8	297.4	1168.8	132.0	1224.7	113.2
DHEA[a]	509.8	53.2	982.7	137.1	1354.6	227.0	1712.2	51.8	1427.5	246.1	1722.0	163.2
Oestrone[a]	9.3	0.9	14.0	2.1	12.2	1.3	35.4	12.4	25.5	2.5	39.1	3.0
Oestradiol[a]	17.9	0.9	16.7	1.6	24.8	2.1	35.6	6.6	37.1	2.1	38.8	2.4
Androstenedione[a]	168.5	14.8	290.8	60.9	302.3	32.1	425.6	51.8	537.7	38.7	868.6	56.2
Progesterone[a]	100.3	7.8	176.2	34.3	121.8	13.8	190.6	37.8	139.2	11.5	169.6	14.9

Notes:
[a] (pg ml^{-1})
[b] (ng ml^{-1})

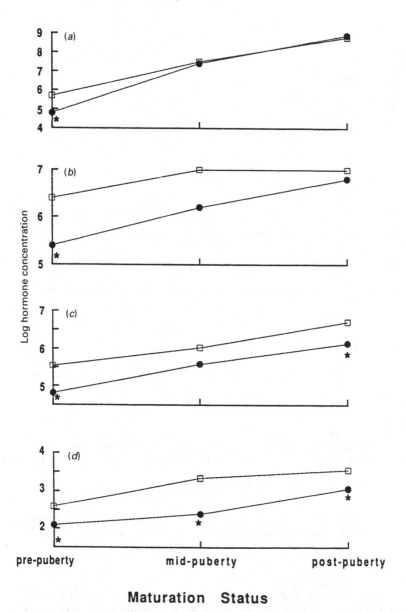

Maturation Status

Fig. 4.5 Urban–rural differences in steroid hormones by maturation status for Bundi males. Means for the natural logarithm concentrations of (*a*) testosterone (pg ml^{-1}), (*b*) DHEAS (ng ml^{-1}), (*c*) androstenedione (pg ml^{-1}) and (*d*) oestradiol (pg ml^{-1}) by maturation status (pre-puberty: testosterone < 1000 pg ml^{-1}; mid-puberty: $1000 \leq$ testosterone ≤ 3000; and post-puberty: testosterone > 3000) showing significant urban–rural differences in androgens and oestrogens. (Symbols: as fig. 4.1.)

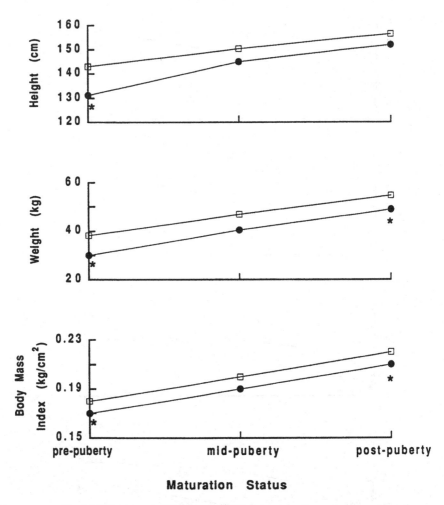

Fig. 4.6 Urban–rural differences in anthropometrics by maturation status for Bundi males. Means for height, weight and body mass index by maturation status, showing significant urban–rural differences in body size. (Symbols: as Fig. 4.1.)

For females (Table 4.3), oestrone, androstenedione and progesterone explained 54% of the variation in height, and, along with DHEAS, they explained 60% of the variation in weight. A slightly different combination of steroids (oestrone, progesterone, DHEAS and oestradiol) were selected for body mass index, accounting for 50% of the variation. Residence was not chosen as a significant predictor of any anthropometric measure. These results are consistent with the hypothesis that the growth of urban and rural females can be explained by endocrine correlates of sexual matura-

Table 4.3. *Stepwise regression of steroid hormones and residence on growth status of Bundi females*

	β	s.e.	F	R^2
Height				
Intercept	95.3			
Log androstenedione	4.0	0.8	15.5	0.45
Log oestrone	4.4	1.1	21.3	0.53
Log progesterone	1.3	0.8	3.0*	0.54
Weight				
Intercept	−14.5			
Log oestrone	4.9	1.3	14.6	0.45
Log androstenedione	2.6	1.0	7.0	0.54
Log progesterone	2.4	0.8	9.2	0.58
Log DHEAS	1.9	1.0	3.2*	0.60
Body mass index				
Intercept	0.1			
Log oestrone	0.01	0.005	4.2	0.39
Log progesterone	0.007	0.003	7.4	0.45
Log DHEAS	0.008	0.003	7.0	0.49
Log oestradiol	0.009	0.005	3.45*	0.50

*, marginally significant $(0.05 \le p \le 0.10)$.

tion, and that once these biological relationships are accounted for, there are no further urban–rural differences in growth status.

For males (Table 4.4), a different pattern emerges. For each anthropometric indicator, testosterone entered the equation first, accounting for the largest portion of the variation in body size. Oestrone and/or oestradiol also were found to be significant predictors of growth status, along with androstenedione for height and weight. Furthermore, urban residence explained a small (1%), but statistically significant amount of variation in height, weight and body mass index, after the endocrine variables entered the model. These models explained 66–77% of the variation in the growth status of Bundi males.

Discussion

In PNG, urbanisation is a fairly recent phenomenon, but for the Bundi, the positive effect of urban residence on growth had already been observed twenty years ago by Malcolm (1970). The present study, using cross-sectional data collected in 1983–4, affirms this previous finding and provides further insight into the mechanisms that underlie this association.

Table 4.4. *Stepwise regression of steroid hormones and residence on growth status of Bundi males*

	β	s.e.	F	R^2
Height				
Intercept	79.3			
Log testosterone	2.3	0.4	31.0	0.63
Log androstenedione	2.7	0.8	10.4	0.72
Log oestradiol	7.4	1.6	22.4	0.76
Urban residence	3.1	1.3	5.7	0.77
Log DHEAS	1.4	0.7	3.5*	0.77
Weight				
Intercept	−15.1			
Log testosterone	2.0	0.4	29.1	0.61
Log oestradiol	6.8	1.7	16.5	0.72
Log oestrone	2.7	1.1	6.5	0.75
Log androstenedione	2.1	0.7	8.8	0.76
Urban residence	2.9	1.3	4.6	0.77
Body mass index				
Intercept	0.1			
Log testosterone	0.005	0.001	27.5	0.53
Log oestrone	0.008	0.003	8.7	0.63
Log oestradiol	0.014	0.004	10.2	0.65
Urban residence	0.007	0.003	4.0	0.66

*, marginally significant $(0.05 \leq p \leq 0.10)$.

Despite a secular trend in growth and age at menarche over the past 18 years, rural Bundi continue to be at the extreme for worldwide variation in growth and sexual maturation. Urban adolescents are significantly taller and heavier than their rural peers, particularly in mid-adolescence when differences in the timing of the adolescent growth spurt results in disparities in height of 9–10 centimetres.

The endocrinological data presented support the hypothesis that these growth differences are due largely to differences in the timing of sexual maturation. For females, the disappearance of urban–rural differences among the older age groups in height and in steroid secretion levels suggests that increased growth status of urban girls is mediated by earlier maturation and does not result in increased adult body size. For males, a different scenario emerges. Urban boys are taller and heavier than rural boys until early adulthood, and the steroid secretory patterns suggest that earlier sexual maturation can account for these differences. However, when comparisons are made on the basis of testosterone level, a measure of gonadal maturation, significant urban–rural differences in other androgens and oestrogens persist. While testosterone may be a good measure of gonadal maturation, the other steroids may reflect other, more subtle maturational processes, such as adrenal maturation, which affect overall growth and maturation. The regression model demonstrates that, in fact, these other hormones are significantly related to the growth status of Bundi males in addition to testosterone, and that urban residence continues to be a significant predictor of growth along with these other biological variables.

Thus, earlier maturation is the most prominent feature of the growth of urban Bundi adolescents. Sex differences in the growth response to the urban environment suggest that maturation rate is quite ecosensitive in females, and that both maturation and body size are sensitive to the environment in males. However, these biological processes are being inferred from cross-sectional associations. Given the complexity of urban lifestyles and environments, and the factors influencing migration, these findings might be taken as preliminary. Urban Bundi are varied in lifestyle, education and economic status, migration history, and residential arrangements. However, there are certain features of urban life that are common: higher population density, dependency on store-bought foods, reduced energy expenditure, and possibly, changes in the household or familial responsibilities of adolescents in comparison with their peers in the village. The availability and quality of health care in urban centres is significantly superior to rural health care such that the disease burden, especially for pneumonia and other respiratory infections, is lower. Nutritional status is also improved by the consumption of snack foods, such as biscuits, scones or ice cream, which are not easily accessible in the remote rural villages where, in some cases, one must walk several hours to reach a trade store.

Urban and rural Bundi also differ in values and economic strategies (Zimmer, 1985). Many urban and some rural Bundi view investment in their children and the success of their children as a long-term means to secure their future, as opposed to investing their resources in the traditional exchange system common throughout the New Guinea Highlands (Zimmer, 1985, and personal communication). Access to better schools and health care in urban areas is thus a major consideration in urban migration. Further, traditional concepts of growth and health are consistent with the notion that careful and attentive feeding of children, especially with 'greasy' foods such as tinned fish or meat, will promote their growth and health (Zimmer, 1985). Thus, differences in economic investment strategies, attitudes towards children and associated behaviours may contribute to the observed urban–rural differences in growth.

As a final consideration, selective migration may be a possible factor contributing to the observed urban–rural differences in Bundi growth and maturation. Some studies have noted differences between migrants and sedentes (Schall, 1990; Mascie-Taylor, 1984; Bogin & MacVean, 1981; Ramirez & Mueller, 1980; Beaglehole *et al.*, 1978). For example, in the Tokelau Island migrant study, male pre-migrants tended to be taller and heavier than male non-migrants. No differences were found for the females (Prior *et al.*, 1974). Thus, it is possible that taller or earlier-maturing Bundi are more likely to migrate successfully to urban areas and secure employment.

In sum, although location, housing, education level and employment are varied among urban Bundi, there are distinctive contrasts from village life, such as changes in health, diet, activity patterns and values, that are likely to contribute to improved growth in the urban environment. Greater steroid hormone levels and earlier maturation are a prominent feature of this pattern. The effects of modernisation and urbanisation on the pace of child development have been repeatedly observed, but this is the first study of the endocrine parameters that underlie this phenomenon. Anthropometric measures, menarcheal age and endocrine profiles all indicate that urban Bundi experience earlier puberty and more rapid growth with accelerated adrenal and gonadal development. Urban Bundi not only mature more quickly, but are also distinctly taller and heavier during adolescence than are their rural counterparts. Since age is reckoned by biological maturation, and not chronologically, these urban young men and women, to whom traditional markers of entry into adulthood are applied, are eligible for courtship, marriage, and parenthood much earlier than are their peers in the village. The social and demographic consequences of these biological changes are yet to be determined.

Acknowledgements

The authors gratefully acknowledge the assistance of Daina Lae, Meza Gini, Jon Shapiro, Travis Jenkins, Ray Sparks and other staff of the PNG Institute of Medical Research for their assistance in data collection and processing. Support for this research was provided by the PNG Institute of Medical Research, the University of Pennsylvania, and graduate fellowships from the National Science Foundation and the American Association of University Women. Above all, we would like to thank the Bundi people for their cooperation and kindness.

References

Abraham, G. E., Buster, J. E., Lucas, L. A., Corrales, P. C. & Teller, R. C. (1972). Chromatographic separation of steroid hormones for use in radioimmunoassay. *Analytic Letters* **5**, 509–17.

Apter, D. (1980). Serum steroids and pituitary hormones in female puberty: a partly longitudinal study. *Clinical Endocrinology* **12**, 107–20.

Apter, D., Lenko, H. L., Perheentupa, J., Soderholm, A. & Vihko, R. (1982). Subnormal pubertal increases of serum androgens in Turner's syndrome. *Hormone Research* **16**, 164–73.

Beaglehole, R., Eyles, E., Salmon, C. & Prior, I. (1978). Blood pressure in Tokelauan children in two contrasting environments. *American Journal of Epidemiology* **108**(4), 283–8.

Bielicki, T. (1986). Physical growth as a measure of the economic well-being of populations: the twentieth century. In *Human Growth. A Comprehensive Treatise*, vol. 3, second edition (ed. F. Faulkner & J. M. Tanner), pp. 283–306. New York: Plenum Press.

Bielicki, T. & Welon, Z. (1982). Growth data as indicators of social inequalities: The case of Poland. *Yearbook of Physical Anthropology* **25**, 153–67.

Bing, C., Xu, S. E., Zhang, G. D. & Wang, W. Y. (1988). Serum dehydroepiandrosterone sulphate and pubertal development in Chinese girls. *Annals of Human Biology* **15**(6), 421–9.

Blizzard, R. M., Thompson, R. G., Baghdassarian, A., Kowarski, A., Migeon, C. J. & Rodriguez, A. (1974). The interrelationship of steroids, growth hormone and other hormones on pubertal growth. In *Control of the Onset of Puberty* (ed. M. M. Grumbach, G. D. Grave & F. E. Mayer), pp. 342–59. New York: John Wiley & Sons.

Boepple, P. A., Mansfield, M. J., Link, K., Crawford, J. D., Crigler, J. F., Kushner, D. C., Blizzard, R. M. & Crowley, W. F. (1988). Impact of sex steroids and their suppression on skeletal growth and maturation. *American Journal of Physiology* **255**(4 part 1), E559–66.

Bogin, B. (1988). Rural-to-urban migration. In *Biological Aspects of Human Migration* (ed. C. G. N. Mascie-Taylor & G. W. Lasker), pp. 90–129. Cambridge University Press.

Bogin, B. & MacVean, R. B. (1981). Bio-social effects of migration on the development of families and children in Guatemala. *American Journal of Public Health* **71**, 1373–7.

Caruso-Nicoletti, M., Cassorla, F., Skerda, M., Ross, J. L., Loriaux, D. L. & Cutler, G. B. (1985). Short term, low dose estradiol accelerates ulnar growth in boys. *Journal of Clinical Endocrinology and Metabolism* **61**, 896–8.

Harrison, G. A., Tanner, J. M., Pilbeam, D. R. & Basker, P. T. (1988). *Human Biology. An Introduction to Human Evolution, Variation, Growth, and Adaptability*. Third Edition. Oxford University Press.

Jenkins, C., Heywood, P. & Zemel, B. (1987). Secular changes in growth in Bundi. In *Rural Health Services in Papua New Guinea* (ed. P. Heywood & B. Hudson), pp. 77–88. P.N.G. Department of Health Monograph No. 5.

Jenkins, C., Smith, T. & Heywood, P. (1990). Ancient diversity and contemporary change in the growth patterns of Papua New Guinea children. *American Journal of Physical Anthropology* **81**(2), 244.

Johnston, F. E., Low, S. M., de Baessa, Y. & MacVean, R. B. (1985). Growth status of disadvantaged urban Guatemalan children of a resettled community. *American Journal of Physical Anthropology* **68**, 215–24.

Kirschner, M. A., Sinhamahapatra, S., Zucker, I. R., Loriaux, L. & Nieschlag, E. (1973). The production, origin and role of dehydroepiandrosterone and androstenediol as androgen prehormones in hirsute women. *Journal of Clinical Endocrinology and Metabolism* **37**, 183–9.

Laska-Meirzejewska, T., Milicer, H. & Piechacek, H. (1982). Age at menarche and its secular trend in urban and rural girls in Poland. *Annals of Human Biology* **9**(3), 227–33.

Laue, L., Kenigsberg, D., Pescovitz, O. H., Hench, K. D., Barnes, K. M., Loriaux, D. L. & Cutler, G. B. (1989). Treatment of familial male precocious puberty with spironolactone and testolactone. *New England Journal of Medicine* **320**(8), 496–502.

Lee, P. A. & Migeon, C. J. (1975). Puberty in boys: correlations of plasma levels of gonadotropins (LH, FSH), androgens (testosterone, androstenedione, dehydroepiandrosterone and its sulfate), estrogens (estrone and estradiol) and progestins (progesterone and 17-hydroxyprogesterone). *Journal of Clinical Endocrinology and Metabolism* **41**, 556–62.

Lee, P. A., Xenakis, T., Winer, J. & Matsenbaugh, S. (1976). Puberty in girls: correlation of serum levels of gonadotropins, prolactin, androgens, estrogens, and progestins with physical changes. *Journal of Clinical Endocrinology* **43**, 775–84.

Malcolm, L. A. (1970). *Growth and Development in New Guinea – A Study of the Bundi People of the Madang District.* Institute of Human Biology Monograph Series no. 1.

Malcolm, L. A. (1975). Some biosocial determinants of the growth, health and nutritional status of Papua New Guinea preschool children. In *Biosocial Interrelations in Population Adaptation* (ed. E. S. Watts, F. E. Johnston & G. W. Lasker), pp. 367–75. The Hague: Mouton.

Malina, R. M. (1979). *The Secular Trend.* Monographs of the Society for Research in Human Development **79**(44), nos. 3–4.

Malina, R. M., Himes, J. H., Dutton Stepick, C., Gutierrez Lopez, F. & Buschang, P. H. (1981). Growth of rural and urban children in the Valley of Oaxaca, Mexico. *American Journal of Physical Anthropology* **55** 269–80.

Manlimos, F. S. & Abraham, G. E. (1975). Chromatographic purification of tritiated steroids prior to use in radioimmunoassay. *Analytic Letters* **8**, 403–10.

Mascie-Taylor, C. G. N. (1984). The interaction between geographical and social mobility. In *Migration and Mobility* (ed. A. J. Boyce), pp. 161–78. London: Taylor & Francis.

McNatty, K. P., Smith, D. M., Makris, A., Osathanondh, R. & Ryan, K. J. (1979a). The microenvironment of the human antral follicle: Interrelationships among the steroid levels in antral fluid, the population of granuloma cells, and the status of the oocyte *in vivo* and *in vitro*. *Journal of Clinical Endocrinology and Metabolism* **49**, 851–60.

McNatty, K. P., Makris, A., DeGrazia, C., Osathanondh, R. & Ryan, K. J. (1979b). The production of progesterone, androgens, and estrogens by granulosa cells, thecal tissue, and stromal tissue from human ovaries *in vitro*. *Journal of Clinical Endocrinology and Metabolism* **49**, 687–99.

Nottelman, E. D., Susman, E. J., Dorn, L. D., Inoff-Germain, G., Cutler, G. B., Loriaux, D. L. & Chrouses, G. P. (1987). Developmental processes in early adolescence. Relations among chronologic age, pubertal stage, height, weight, and serum levels of gonadotropins, sex steroids and adrenal androgens. *Journal of Adolescent Health Care* **8**, 246–60.

Preece, M. A., Cameron, N., Donmall, M. C., Dunger, D. B., Holder, A. T., Preece, J. B., Seth, J., Sharp, G. & Taylor, A. M. (1984). The endocrinology of male puberty. In *Human Growth and Development* (ed. J. Borms, R. Hauspie, A. Sand, C. Susanne & M. Hebbelink), pp. 23–37. New York: Plenum Press.

Prior, I., Stanhope, J. M., Evans, J. G. & Salmond, C. E. (1974). The Tokelau Islands Migrant Study. *International Journal of Epidemiology* **3**(3), 225–32.

Raisz, L. G. & Kream, B. E. (1981). Hormonal control of skeletal growth. *Annual Review of Physiology* **43**, 225–38.

Ramirez, M. E. & Mueller, W. H. (1980). The development of obesity and fat patterning in Tokelau children. *Human Biology* **52**(4), 675–87.

Schall, J. (1990). Blood pressure, body size and fat patterns of Manus youth: a migrant study. *American Journal of Physical Anthropology* **81**(2), 291

Sizonenko, P. C. & Paunier, L. (1975). Hormonal changes in puberty III: correlations of plasma dehydroepiandrosterone, testosterone, FSH, and LH with stages of puberty and bone age in normal boys and girls and in patients with Addison's disease or hypogonadism or with premature or late adrenarche. *Journal of Clinical Endocrinology and Metabolism* **41**, 894–904.

Sizonenko, P. C., Paunier, L. & Carmignac, D. (1976). Hormonal changes during puberty. IV. Longitudinal study of adrenal androgen secretions. *Hormone Research* **7**, 288–302.

Tanner, J. M., Whitehouse, R. H., Hughes, P. C. R. & Carter, B. S. (1976). Relative importance of growth hormone and sex steroids for the growth at puberty of trunk length, limb length, and muscle width in growth hormone-deficient children. *Journal of Pediatrics* **89**(6), 1000–8.

Wen, X., Villee, D. B., Ellison, P., Todd, R. & Loring, J. (1985). Effects of adrenocorticotropic hormone, human chorionic gonadotropin, and insulin on steroid production by human adrenocortical carcinoma cells in culture. *Cancer Research* **45**, 3974–8.

Wierman, M. E., Beardsworth, D. E., Crawford, J. D., Crigler, J. F., Mansfield, M. J., Bode, H. H., Boepple, P. A., Kushner, D. C. & Crowley, W. F. (1986). Adrenarche and skeletal maturation during luteinizing hormone releasing hormone analogue suppression of gonadarche. *Journal of Clinical Investigation* **77**, 121–6.

Winter, J. S. D. (1978). Prepubertal and pubertal endocrinology. In *Human Growth*, vol. 2 (ed. F. Faulkner & J. M. Tanner), pp. 183–213. New York: Plenum Press.

Worthman, C. (1987). Interactions of physical maturation and cultural practice in ontogeny: Kikuyu adolescents. *Cultural Anthropology* **2**, 29–38.

Worthman, C. & Whiting, J. W. M. (1987). Social change in sexual behavior and mate selection in a Kikuyu community. *Ethos* **15**, 145–65.

Zachmann, M., Prader, A., Sobel, E. W., Crigler, J. F. Jr, Ritzen, E. M., Atares, M. & Ferrandez, A. (1986). Pubertal growth in patients with androgen insensitivity: indirect evidence for the importance of estrogens in pubertal growth of girls. *Journal of Pediatrics* **108** (5 part 1), 694–7.

Zemel, B. S. (1989). *Dietary Change and Adolescent Growth among the Bundi (Gende-speaking) People of Papua New Guinea.* Ph.D. dissertation in Anthropology, University of Pennsylvania.

Zemel, B. S. & Jenkins, C. (1989). Dietary change and adolescent growth among the Bundi (Gende-speaking) people of Papua New Guinea. *American Journal of Human Biology* 1(6), 709–18.

Zemel, B. S. & Katz, S. H. (1986). The contribution of adrenal and gonadal androgens to the growth in height of adolescent males. *American Journal of Physical Anthropology* **71**, 459–66.

Zhang, X. & Huang, A. (1988). The second national growth and development survey of children in China, 1985: children 0 to 7 years. *Annals of Human Biology* **15**(4), 289–305.

Zimmer, L. 1985). *The Losing Game – Exchange, Migrations, and Inequality among the Gende People of Papua New Guinea.* Ph.D. dissertation in Anthropology, Bryn Mawr College.

Zimmer, L. Housewives, homemakers and household managers. Gende women in town. In *Modern Papua New Guinea Society* (ed. L. Zimmer). Bathurst, Australia: Crawford House Press. (In press.)

5 Nutritionally vulnerable households in the urban slum economy: a case study from Khulna, Bangladesh

J. PRYER

Introduction

It has been established that undernutrition is not randomly distributed within a given population, but is a dimension of poverty (Basta, 1977; Hussain & Lunven, 1987; Lipton, 1983; Nabarro, 1984; Payne, 1987; Rao, 1985; Reichmanheim, 1988; Sukhatme, 1981; UNICEF, 1981). The ability of a household to command sufficient food resources is primarily dependent upon social and economic variables such as assets, employment and income.

In urban areas, households are often characterised by a dependence upon the market, not only for employment and food, but also for other basic needs. Several studies conducted in urban areas of developing countries have indicated undernutrition in children to be associated with low effective income (Aguillon et al., 1985; Azevedo, 1989; Mazur & Sanders, 1988; Murillo-Gonzalez, 1983; Victora et al., 1986; Wray & Aguirre, 1969), low levels of food expenditure (Wray & Aguirre, 1969), low levels of ownership of consumer durables (Azevedo, 1989; Pickering et al., 1985), or low levels on composite socioeconomic scores (Christiansen et al., 1975; Greiner & Latham, 1981; Zeitlin et al., 1978).

Rarely have such studies attempted to identify the types of urban livelihoods associated with low income and undernutrition. The aim of this paper is to describe patterns of household economic livelihood in one inner-city slum in the city of Khulna, Bangladesh, and to identify those patterns of livelihood that were most closely associated with nutritional vulnerability among both children and adults.

The paper is based upon a detailed study, which utilised both quantitative cross-sectional household survey techniques and extended qualitative case study methods. The combination of these two techniques was purposely chosen in an attempt to overcome difficulties of discerning processes from cross-sectional data (Strickland, 1988).

The paper is divided into five sections. Firstly, a brief background to the urban environment in Khulna is described. This is followed by the study's conceptual framework. On the basis of this conceptual framework and using the multivariate technique of cluster analysis, five livelihood groups were identified in the slum; their characteristics are presented in the third section. Finally, the patterns of livelihood most closely associated with nutritional vulnerability in both children and adults are described, and conclusions are then drawn.

Medja Para slum: the environmental context

Khulna is the second seaport and third city of Bangladesh, with a population in 1986 of around 750 000. The city has a modest and diversified industrial base and has historically been an important regional trading centre. The city has been growing rapidly in recent decades and this growth is expected to continue; the estimated population in the year 2000, with a moderate growth rate, is 2 million (Islam & Haq, 1981). Apart from natural increase in the urban population, the impetus behind this growth has been industrialisation and migrant influxes after political disturbances (the partition of India and Pakistan in 1947; communal riots in Calcutta in 1965; and the liberation war of Bangladesh in 1971). In addition, immigrants come from surrounding rural areas as a result of the ongoing process of rural pauperisation (CUS, 1974; Islam & Haq, 1981). The result has been a mushroom growth of both legal and illegal slums; these are officially estimated to contain around 50% of the city population (World Bank, 1981).

Medja Para (a pseudonym), the site of current research, is an established inner-city slum with a population of over 2200. Environmental conditions are abhorrent, being insanitary and overcrowded; the population density in the slum was 1200 persons per acre (2963 per hectare) in 1986–7. Nutritional and environmental diseases abound in these conditions. Cross-sectional surveys in Medja Para between September and December 1986 indicated that 7% of children under five years were severely undernourished and 43% moderately undernourished using weight-for-age as an indicator. Relatively high levels of ill-health and undernutrition were not solely confined to young children: 43% of mothers and 42% of fathers had a Body Mass Index (BMI)[1] below 18, which is considered to be potentially indicative of adult undernutrition (James et al., 1988).

Trade and labour of household members were the most important sources of income in the slum. The slum labour market was highly differentiated by gender, reflecting the system of severe social control over women in Bangladesh (Kabeer, 1988). Employment opportunities for poor and uneducated women were scarce and restricted. Despite an initial impression of homogeneous poverty there was a high degree of inequality within the slum. The Gini coefficient for asset ownership was very high (0.77), indicating extreme inequality in concentration of asset ownership; the top 10% of households owned 70% of all assets, whilst the bottom 10% owned only 0.07%. Around 50% of households were below a slum based food poverty line[2], and 25% of households were 'ultra-poor' (Lipton, 1983).

The study's conceptual framework

To date, most urban nutrition studies with a socioeconomic focus have tended to lack a theoretical contextualisation in terms of the central processes that characterise and differentiate urban households and their forms of insertion into the wider urban economy. Fig. 5.1 presents a conceptual framework for the analysis of variations in material conditions of urban slum households and the nutritional outcomes of household reproduction by focussing on the inter-related processes of production, circulation and reproduction. The model is an integration and adaptation of two published frameworks: the urban labour status framework developed by Bromley & Gerry (1979), and a model of the organisation of peasant households developed by Deere & de Janvry (1979).

In Fig. 5.1, a household is considered to be endowed with a set of resources such as labour power, the means of work and raw materials. This endowment conditions the array of economic choices available to it on the urban labour market. Income is generated through any combination of the trade and wage labour production processes or from home production. After deduction of the various costs of production, the net income so derived permits the purchase of the means of consumption for the reproduction of the household.

Twelve sets of key indicator variables can be identified within this framework as a means of analysing empirically the organisation of households:

1 the stocks of the means of production;
2 household demographic characteristics and the division of productive labour;
3 the home production process;
4 the self-employed and petty commodity production process;
5 the dependent labour production process;
6 the disguised wage labour production process;
7 the short-term wage labour production process;
8 the permanent wage labour production process;
9 formation of household income;
10 credit and debt;

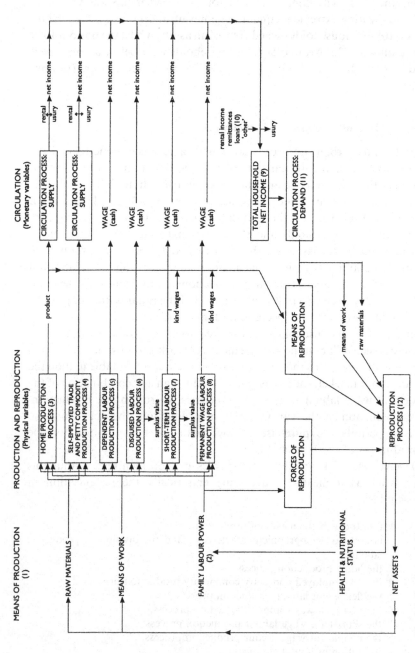

Fig. 5.1 Conceptual framework of processes operating at a household level. Adapted from Bromley & Gerry (1979) and Deere & De Janvry (1979).

11 effective demand – disposal of income;
12 reproduction of the consumption unit and the net asset position of households.

In Medja Para slum, the home production process, one of the most important processes through which income is generated in rural areas, is of minor importance. No urban agriculture existed, although the rearing of small livestock was undertaken in some households. However, in other urban contexts, urban agriculture and livestock rearing have been found to be a significant contributor to total household food consumption (Evers, 1981).

The conceptual framework is outlined below and has been discussed in more detail elsewhere (Pryer, 1990).

Production and circulation

(i) Means of production
These are located in the left-hand column of Fig. 5.1 and include raw materials (land and water), means of work (tools, animals, equipment and fuel) and household labour power (the number, age and gender of household members and the level of skill or education).

(ii) Processes of production and circulation
Household labour power may be used in combination with any of the complementary resources at its disposal, in trade or petty commodity production, in disguised or dependent labour, in short-term or permanent labour, or in the home production process, Households participating in petty commodity production or home production generate a gross product, which is either sold onto the market (the circulation process) or retained for home consumption. The sale of commodities and the wages received from wage labour constitute a gross monetary income. This income after the deduction of the production costs generates a net income, which permits the purchase of the means of consumption for the reproduction of the household (family labour power) and the means of work for replacement and net investment. The processes of production and circulation are shown in the central and right-hand columns of Fig. 5.1.

(iii) Surplus extraction
The social relations of production into which households participate are defined in the manner in which surplus is appropriated. Empirically, it is necessary to identify the mechanisms through which surplus appropriation occurs in order to analyse the relations of production into which

households enter. Deere & de Janvry (1979) identified seven mechanisms of surplus extraction: three operate via rents that result from private appropriation of land (rents in labour services, kind and cash); three operate via markets (for labour, products and money); and one operates via the State (taxes). Of these, five were seen to operate in the urban slum in Khulna: two via markets (labour and money), two via private appropriation of land (rents in cash and kind), and one via the State (taxes).

The process of reproduction

The process of human reproduction or the reproduction of labour include the forces of reproduction, and the means of consumption and reproductive work (Bryceson, 1980; Edholm *et al.*, 1977; Mackintosh, 1981; McDonough and Harrison 1978). The forces of reproduction include household size, age and gender structure and the level of skill or education and health and nutritional status of household members. The means of consumption include the quantity of food produced by the household and the level of income derived from the processes of production and circulation with which food and health care can be purchased. The means of reproductive work include housing, water and fuel supply, sanitation, domestic equipment and access to health care supplies and technology (for example, medicines and contraceptives).

In this study no attempt has been made to describe the complex process of reproduction in any depth as it was not possible within the study context to collect data on domestic time allocation, decision-making, demographic reproductive strategies etc. Rather, the 'potential' of households to reproduce on a daily basis will be assessed by comparing household income against locally derived food poverty lines and household level daily food supply.

Outcomes of the processes of production and reproduction

An important dimension of this model is that health and nutritional status of household members is seen simultaneously as an outcome of and an input into the processes of production and reproduction. A nutritional profile of children and adults who come from households with different patterns of livelihood will thus be described. Information on incapacitating ill-health amongst economically active household members was also used to ascertain the constraint that ill-health placed upon the pattern and level of livelihood pursued by households during the study period. Other material outcomes of the processes of household production and reproduction, which are of lesser concern in this study, include net assets for investment, working capital, tools and equipment.

Identification of patterns of livelihood within the slum

Utilising the twelve sets of key indicator variables identified in the study conceptual framework (refer to Fig. 5.1), the multivariate statistical technique of cluster analysis was applied to the Medja Para household data in an attempt to identify relatively homogeneous groups within the slum which shared common patterns and levels of livelihood (Pryer, 1990). Cluster analysis as a technique aims to identify relatively homogeneous groups of cases based upon selected attributes. The technique aims to maximise within-group homogeneity and at the same time to maximise between-group heterogeneity. As a multivariate technique it has a major advantage over univariate methods of classification in that it is not dependent upon a single variable to classify households, but rather is based upon a multidimensional view of livelihood strategies (Everitt, 1980).

Five clearly defined groups emerged from these analyses (Pryer, 1990). Ninety-eight per cent of households could be described in terms of four groups, while the remaining two per cent comprised a fifth group of only four households. The degree of dissimilarity between this small group of four households and the rest of the study population was greater than the degree of dissimilarity between the rest of the groups. Although this group was quantitatively insignificant, it was important in terms of understanding the social structure of the slum. Table 5.1 presents a summary of economic and demographic indicators by cluster livelihood group. A detailed analysis of each group has been reported elsewhere (Pryer, 1990).

Cluster group 1(C1) ($n = 4$) was the richest group within the slum. They were the largest and most politically powerful slum landlords and were well endowed with a valuable and diversified asset base. Their main livelihood strategy was based upon the rental of slum property, agriculture and trade. High levels of business indebtedness, from formal bank sources at low rates of interest, indicated further investment in their asset base. This group has the highest mean household size, coupled with a high economic dependency ratio. Labour participation was lowest in the slum overall, and women and children were not labour market participants. This group had the highest potential to reproduce. Incomes were four times the slum poverty line, and households had the highest food energy supplies and enjoyed a diversified diet.

Cluster group 2 ($n = 28$) was another relatively rich group within Medja Para slum, although to a lesser extent than C1. Income and assets were well

Table 5.1. *Summary characteristics of cluster livelihood groups*

Figures given are cluster livelihood group means; standard deviations in parentheses.

Variable	C1 (n=4)	C2 (n=28)	C3 (n=56)	C4 (n=71)	C5 (n=49)	F^b	P
Assets							
Total assets[a]	177.7	48.9	6.6	1.6	0.6	64.56	0.00
('000 Tk/cu)	(85.1)	(11.4)	(11.6)	(2.5)	(1.3)		
Fixed assets (as %	88.2	48.3	44.8	37.5	31.6	4.9	0.00
of total assets)	(15.5)	(33.9)	(37.8)	(40.2)	(40.3)		
Non-fixed assets (%	11.8	51.7	55.2	62.1	68.2	2.13	0.07
of total assets	(15.5)	(33.9)	(37.8)	(40.7)	(43.9)		
Demographic indicators							
Household size	7.8	6.5	5.6	5.6	6.2	2.71	0.03
	(3.6)	(3.3)	(1.7)	(1.6)	(1.9)		
Economic dependency	5.1	3.3	3.5	3.1	2.2	9.98	0.00
ratio (D/E)	(0.9)	(1.6)	(1.5)	(1.6)	(1.7)		
% Female headed or supported households	0	0	2	1	34	—	—
Labour participation (days/month)							
Fathers	26.3	26.2	24.2	24.3	14.1	15.83	0.00
	(2.5)	(3.4)	(7.0)	(7.1)	(11.6)		
Mothers	0	1.1	0.9	3.8	13.4	43.6^c	0.00
		(4.3)	(4.8)	(8.8)	(14.1)		
Children	0	1.1	0.7	1.9	11.1	41.1^c	0.00
		(5.7)	(4.0)	(8.6)	(15.2)		
Days lost due to	0	0.4	0.8	1.6	7.1	32.1^c	0.00
sickness		(1.5)	(1.9)	(5.1)	(10.5)		
% of total income from							
Cash labour	0	31.2	17.0	77.7	75.7	34.01	0.00
		(41.1)	(33.7)	(32.5)	(32.5)		
Kind labour	0	1.6	0.9	1.3	12.6	12.52	0.00
		(5.1)	(3.9)	(4.1)	(19.8)		
Trade	26.7	56.7	70.7	14.7	14.0	28.26	0.00
	(18.0)	(41.7)	(39.4)	(31.6)	(26.5)		
Agriculture	34.0	8.7	3.6	1.0	1.0	12.16	0.00
	(35.2)	(8.0)	(10.3)	(5.6)	(8.0)		
Rent	35.0	3.7	2.1	1.4	0.7	13.02	0.00
	(27.7)	(7.8)	(6.9)	(11.7)	(5.3)		
Effective demand							
% income	32.8	57.4	89.9	91.6	94.7	7.66	0.00
on food	(12.3)	(22.3)	(30.9)	(41.9)	(41.9)		
Debt							
Productive loans	742.8	15.9	6.4	1.6	1.0	7.8	0.00
('00 Tk/cu)	(856.5)	(44.3)	(12.4)	(2.7)	(2.3)		

Table 5.1. (*cont.*)

Figures given are cluster livelihood group means; standard deviations in parentheses.

Variable	C1 (n=4)	C2 (n=28)	C3 (n=56)	C4 (n=71)	C5 (n=49)	F^b	P
Consumption loans (Tk/cu)	0.0	67.8 (239.4)	109.3 (248.7)	108.2 (147.5)	280.5 (376.7)	9.47	0.00
Reproduction of the consumption unit							
Income as % of SPL^d	460.3 (137.1)	228.3 (134.9)	119.3 (56.5)	87.7 (36.2)	64.8 (28.2)	48.0	0.00
Net assets ('000 Tk/cu)	103.4 (145.6)	17.2 (11.0)	5.9 (11.5)	1.3 (2.5)	0.2 (1.3)	39.6	0.00

Notes:
^{a} Tk/cu: Tk, Bangladesh currency (46 Taka = one pound sterling in December, 1986); cu, Nutritional Consumption Unit.
^{b} One-way analysis of variance F statistic. The distribution of values for most socioeconomic variables was skewed. The data was transformed using natural logarithms. The means and standard deviations in Table 5.1 refer to untransformed values, whereas the F statistic and P values were obtained after logarithmic transformation.
^{c} Non-parametric Kruskal–Wallis one-way ANOVA; H statistic corrected for ties.
^{d} SPL, slum poverty line.

above population means. The predominant C2 livelihood strategy was trade, although income from salaried and 'permanent' wage work were also important. A quarter of Cluster group 2 householders were slum landlords, but their landholdings were small and rental income as well as agricultural income was low. C2 contained the lowest proportion of indebted households; loans incurred were mainly for business purposes. C2 households had a lower access to formal credit than C1 households, and overall interest rates incurred were higher, although still lower than in the poorest livelihood groups. The mean C2 household size and economic dependency ratio were well above slum means, but lower than in C1. As in C1, labour participation was almost exclusively confined to adult men.

In marked contrast to these two relatively rich groups, Cluster group 5 (n=49) was potentially the most vulnerable within the slum. They had lower mean values for income, food supply, assets and net assets than any other group. C5 also had a relatively large mean household size, but unlike the relatively richer C1 and C2 households, this was coupled with a low mean economic dependency ratio. C5 was characterised by high levels of chronically ill adult males who were incapacitated from wage work. Female-headed and supported households were also concentrated in this group. C5 pursued a livelihood strategy based upon unskilled labour and high levels of female and child labour participation. Consumption indebtedness from the informal credit market, at usurious rates of interest, was also a characteristic feature. Despite high levels of indebtedness and a high proportion of income being allocated to food, C5 households had the least potential to reproduce and could be considered potentially the most nutritionally vulnerable group within the slum.

Cluster groups 3 and 4 ($n = 56$ and $n = 71$, respectively), were juxtaposed between these extremes. C3 consisted predominantly of small traders and according to a range of economic and demographic indicators could be considered the 'average' slum household. In terms of livelihood strategy C4 relied predominantly upon the labour of household members, a similar livelihood strategy to C5. However, where C4 and C5 differed most markedly was in (a) the level of male, female ane child labour participation; (b) demographic characteristics, particularly household structure and size and the economic dependency ratio; (c) the extent of earner-incapacitating ill-health; (d) the extent of consumption indebtedness; and (e) the intensity of poverty, as indicated by total expenditure, ownership of assets and net assets. In all these aspects C4 was either similar to C3 or juxtaposed between C3 and C5.

Outcomes of the processes of production and reproduction

Cluster livelihood differentials in child anthropometric status were highly statistically significant for all three anthropometric indicators. The degree of statistical significance, as indicated by the F statistic, was strongest for weight-for-age (WA) and weakest for weight-for-height (WH) (Table 5.2). Although a negative C1–C5 gradient was evident in the mean WA, HA and WH anthropometric scores, the differential between the anthropometric means of C5 and the means of the other livelihood groups were the most pronounced (Table 5.2). This observation was confirmed by the Scheffe Test, which only identified C5 by anthropometric means as being statistically different from that of any other livelihood group.

In summary, of the three anthropometric indicators, weight-for-age (a composite indicator of both weight-for-height and height-for-age) was the most strongly correlated with livelihood group: 68% of all severely weight-for-age undernourished children (i.e. below 60% of the NCHS median) were located in C5. The relative risk of a severely weight-for-age or severely height-for-age undernourished children coming from a C5 household was 7.5 times and 3.9 times greater, respectively, than from a non-C5 household.

Marked livelihood group differences were also evident in the distribution of parental anthropometric status. Table 5.3 shows a significantly negative

Table 5.2. *Child anthropometric status by gender and livelihood group*

Figures are for children aged 0–60 months and are cluster livelihood group means (standard deviation in parentheses).

Variable	C1+2	C3	C4	C5	F	P
All children (*n*)	50	85	109	69		
Weight-for-age (%)	77.9	77.8	75.0	69.7	11.4	0.00
	(9.8)	(8.9)	(9.9)	(11.1)		
Height-for-age (%)	91.4	90.7	89.9	87.9	6.98	0.00
	(3.8)	(3.6)	(4.6)	(6.5)		
Weight-for-height						
(%)	92.3	92.9	91.5	87.4	4.3	0.00
	(9.7)	(7.8)	(10.1)	(9.2)		

Table 5.3. *Parental anthropometric status by livelihood group*

Variable	C1+2	C3	C4	C5	F	P
Fathers (*n*)	34	55	70	40		
BMI	20.4	19.1	18.8	17.5	9.00	0.00
	(2.4)	(2.8)	(2.2)	(1.6)		
Non-pregnant						
mothers (*n*)	31	47	67	41		
BMI	20.1	19.0	18.7	17.3	8.05	0.00
	(3.3)	(2.3)	(2.3)	(1.4)		

Notes: BMI = Body Mass Index (weight in kilogrammes divided by the square of height in metres)

C1+2 – C5 gradient in mean BMI for both fathers and non-pregnant mothers. The mean BMI for non-pregnant mothers and fathers in the poorest C5 was 17.4 and 17.5, respectively, compared with 20.1 and 20.4 for non-pregnant mothers and fathers, respectively, in the richest C1+2. Livelihood group differences in mean BMI of both fathers and mothers were reflected in the frequency distribution in the different BMI categories. For example, 45% of C5 fathers had a BMI below 17, compared with 9% in C1+2.

Summary

Most urban studies with a socioeconomic status relate household socioeconomic indicators, such as income or expenditure, directly to anthropometric indicators without first analysing the social processes through which income is generated. The aim of this paper was to describe patterns of household economic livelihood in an inner-city slum in the city of Khulna, Bangladesh, and to identify those patterns of livelihood which were most closely associated with nutritional vulnerability among both children and adults.

A conceptual framework was presented for an analysis of variations in material conditions of urban slum households and the nutritional outcomes of household reproduction. Twelve sets of key indicator variables were identified as a means of analysing empirically the organisation of urban households. These variables formed the basis within which five relatively homogeneous groups of households were identified by using the multivariate statistical technique of cluster analysis. The livelihood group that was identified as the most nutritionally vulnerable within the slum had the lowest mean values for income, assets and net assets of any group, and also had high levels of consumption indebtedness from informal credit markets at usurious rates of interest. The group had a high mean household size in conjunction with a low dependency ratio, reflecting high levels of female and child labour participation. C5 was characterised by a high proportion of female-headed and -supported households, and a high proportion of households with earners incapacitated from wage work owing to ill-health. This group had the least potential to reproduce and were the most nutritionally vulnerable within the slum. All the key indicator variables that differentiated C5 households from other cluster livelihood groups were subsequently found to be highly statistically associated with the weight-for-age status of children within the household (Pryer, 1990).

Cluster analysis has seldom been used in social nutrition. Its successful application here indicates its potential value either in empirically differentiating households or alternatively in validating the efficacy of *ex ante* models of household classification.

Acknowledgements

Thanks are due to the Overseas Development Administration (UK), Oxfam, and Save the Children Fund (UK) for funding the fieldwork and analysis on which this chapter is based; to Drs Barbara Harriss and Edward Clay for academic supervision; and to Dr Simon Stickland for comments on this chapter. Drs Kabir and Khanum and Mr Ghias Uddin of the Save the children Fund (UK) provided invaluable institutional support in Bangladesh.

Notes
[1] Body Mass Index (BMI) is body mass in kilograms divided by the square of height in metres.

² The slum-based food poverty line was based on a weighted basket of the foods n ost commonly purchased by the slum households. The value of the food poverty line in 1986–7 was 291 Taka per consumption unit per month. A consumption unit was taken to be the WHO energy requirements of a moderately active man; 46 Taka = one pound sterling in December 1986.

References

Aguillon, D.B., Caedo, M.M., Arnold, J.C. & Engel, R.W. (1985). The relationship of family characteristics to the nutritional status of pre-school children. *Food and Nutrition Bulletin* **4**(4), 5–11.

Azevedo, I.C.B. (1989). *Child growth and adult body size in a poor region of north-east Brazil: anthropometry as an index of risk in individuals and households*. Ph.D. Thesis, London School of Hygiene and Tropical Medicine.

Basta, S.S. (1977). Nutrition and health in low income urban areas of the third world. *Ecology of Food and Nutrition* **6**, 113–24.

Bromley, R. & Gerry, C. (1979). *Casual Work and Poverty in Third World Cities*. Chichester: John Wiley and Sons.

Bryceson, D. (1980). Proleterianisation of Tanzanian women. *Review of African Political Economy* **17**, 4–27.

Centre for Urban Studies (1974). *Squatters in Bangladesh Cities: a Survey of Squatters in Dacca, Chittagong and Khulna*. Dhaka, Bangladesh: Urban Development Directorate, Ministry of Public Works, and the Centre for Urban Studies, University of Dhaka.

Christiansen, N., Mora, J.O. & Herrera, M.G. (1975). Family social characteristics related to physical growth of young children. *British Journal of Preventative and Social Medicine* **29**, 121–30.

Deere, C.D. & de Janvry, A. (1979). A conceptual framework for the empirical analysis of peasants. *American Journal of Agricultural Economics November, 1979*, pp. 601–11.

Edholm, F., Harriss, O. & Young, K. (1977). Conceptualising women. *Critique of Anthropology* **3**(9/10), 101–30.

Everitt, B. (1980). *Cluster Analysis*, 2nd edn. London: Heinemann Educational Books.

Evers, H.D. (1981). The contribution of urban subsistence production to incomes in Jakarta. *Bulletin of Indonesian Economic Studies* **17**(2), 89–96.

Greiner, T. & Latham, M.C. (1981). Factors associated with nutritional status among young children in St. Vincent. *Ecology of Food and Nutrition* **10**, 135–41.

Hussain, A.M. & Lunven, P. (1987). Urbanisation and hunger in the cities. *Food and Nutrition Bulletin* **9**(4), 50–61.

Islam, N. & Haq, Z.S. (1981). *Population, migration characteristics of Khulna city*. Dhaka, Bangladesh: Centre for Urban Studies, University of Dhaka.

James, W.P.T., Ferro-Luzzi, A. & Waterlow, J.C. (1988). Definition of chronic energy deficiency in adults. *European Journal of Clinical Nutrition* **42**, 969–81.

Kabeer, N. (1988). Subordination and struggle: women in Bangladesh. *New Left Review* **168**, 95–121.

Lipton, M. (1983). Poverty, nutrition and hunger. *World Bank Staff Working Paper* No. 597. Washington, D.C.: World Bank.

74   *J. Pryer*

Mackintosh, M. (1981). Gender and economics: the sexual division of labour and the subordination of women. In *Of Marriage and the Market* (ed. K. Young, C. Wolkowitz & R. McCullagh), pp. 1–15. London: CSE Books.

Mazur, R. & Sanders, D. (1988). Socio-economic factors associated with child health and nutrition in peri-urban Zimbabwe. *Ecology of Food and Nutrition* **22**, 19–35.

McDonough, R. & Harrison, R. (1978). Patriarchy and relations of production. In *Feminism and Materialism* (ed. A. Kuhn & A. M. Wolpe), pp. 11–41. London: Routledge and Kegan Paul.

Murillo-Gonzalez, S. T. (1983). *The effects of social factors on the nutritional status of children in urban Costa Rica.* Ph.D. thesis, London School of Hygiene and Tropical Medicine.

Nabarro, D. (1984). Social, economic, health and environmental determinants of nutritional status: a case study from Nepal. *Food and Nutritional Bulletin* **6**(1), 18–32.

Payne, P. R. (1987). *Undernutrition measurement and implications.* Paper prepared for WIDER conference on poverty, undernutrition and living standards, Helsinki, 27–31 July 1987.

Pickering, H., Hayes, R. J., Dunn, D. T. & Tomkins, A. M. (1985). *Socio-economic factors associated with child growth in a peri-urban community in Gambia.* Dept. of Human Nutrition, London School of Hygiene and Tropical Medicine (mimeograph).

Pryer, J. (1990). *Socio-economic and environmental aspects of undernutrition and ill-health in an urban slum in Bangladesh.* Ph.D. thesis, University of London.

Rao, K. S. J. (1985). Urban nutrition in India. *Bulletin of the Nutrition Foundation of India* **16**, no. 28.

Reichmanheim, M. E. (1988). *Child health in an urban context: risk factors in a squatter settlement of Rio de Janeiro.* Ph.D. thesis, London School of Hygiene and Tropical Medicine.

Strickland, S. (1988). *Workshop on household profiles: summary of discussion points.* Dept. of Human Nutrition, London School of Hygiene and Tropical Medicine (mimeograph).

Sukhatme, P. V. (1981). *Relationship between malnutrition and poverty.* Paper presented at First National Conference on Social Sciences. National Council of Applied Economic Research, New Delhi.

UNICEF (1981). Socio-economic characteristics of functional groups of malnourished populations in Costa Rica. *Social Statistics Bulletin, East African Regional Office (UNICEF)* **4**, no. 4.

Victora, C. G., Vaughn, P., Kirkwood, B. R., Martines, J. C. & Barcelos, L. B. (1986). Risk factors for malnutrition in Brazilian children: the role of social and environmental variables. *Bulletin of the World Health Organisation* **2**, 299–309.

World Bank (1981). *Urban sector memorandum.* World Bank Report No. 3422-BD, Dhaka, May 1981.

Wray, J. D. & Aguirre, A. (1969). Protein calorie malnutrition in Candelaria, Colombia. 1. Prevalence, social and demographic causal factors. *Journal of Paediatrics* **15**, 76.

Zeitlin, M., Masangkay, Z., Consolacion, M. & Nass, M. (1978). Breastfeeding and nutritional status in depressed urban areas of greater Manilla, Philippines. *Ecology of Food and Nutrition* **7**, 103–13.

6 Urban–rural differences in growth and diarrhoeal morbidity of Filipino infants

L. S. ADAIR, J. VANDERSLICE AND N. ZOHOORI

Introduction

Numerous reports that present data on child growth and health outcomes around the world highlight urban–rural differences. Many of the classic growth studies reported in Tanner & Eveleth (1976) show that, in general, children in urban areas tend to be healthier, taller and heavier than children in surrounding rural areas. Similarly, data from the FAO's 1985 Fifth World Survey (presented in Fig. 1 of Keller, 1988) show consistent urban–rural differences in the prevalence of stunting and wasting among 0–5 year old children (see Fig. 6.1). The better health status of urban children is often attributed to a regular supply of goods, health and sanitation services, education, and medical facilities, associated with the urban environment.

If we look at large urban centres of the developing world over the past two decades, however, a different picture begins to emerge (Popkin & Bisgrove, 1988). Urban centres are not necessarily uniformly healthy environments. Population growth in urban centres is proceeding at a very rapid rate: by the year 2000, it is projected that 43% of people in developing countries will live in cities (United Nations, 1980). Concomitant with rapid population growth, due to both migration and natural increase, is a dramatic shift in demographic and socioeconomic composition of urban areas. Approximately one third of the world's poor now live in cities (Churchill, 1980), mostly in makeshift squatter settlements that share problems of poverty, overcrowding, poor environmental conditions and psychological stress. In more recent studies comparing growth of children in urban slums with that of children in rural areas, it is clear that the urban 'advantage' disappears: urban slum children suffer higher rates of morbidity and attain growth status comparable to that of rural children (cf. Underwood & Margetts, 1987; Nutrition Foundation of India, 1988).

When researchers look at urban–rural differences, their underlying assumption is that there is a set of characteristics that jointly occur in one

75

76   *L. S. Adair* et al.

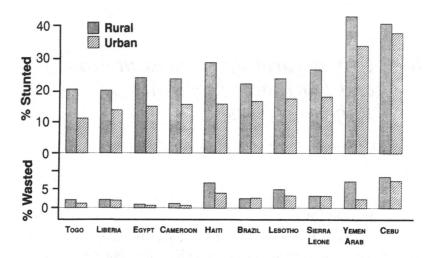

Fig. 6.1 Percentages of stunted and wasted children in the age groups 0–5 years, by urban and rural residence in selected countries. Modified from Keller (1988) with the addition of data on 0–1-year-old children from Cebu, Philippines.

environment and not the other, and further that some or all of these characteristics are important determinants of health and growth. However, specific characteristics of urban environments have not been systematically or explicitly linked to health outcomes in many of the studies that have been done. Furthermore, there may be characteristics thought to be associated with an urban environment which, at the household level, are not highly correlated with urban residence (for example, poor household sanitation).

In this paper, we wish to explore characteristics of urban and rural environments that have a significant impact on child health and growth. We do this by using data from a large survey of child health and survival (the Cebu Longitudinal Health and Nutrition Survey, CLHNS) conducted in Metro Cebu, Philippines. The main objective of the CLHNS was to develop a comprehensive model that links underlying community, household and individual factors, intermediate health-related behaviours and health outcomes (growth, diarrhoea and respiratory morbidity) to produce a more complete picture of the determinants of child health. Work by the Cebu Study Team to date has shown consistent urban–rural differences in health outcomes in children and mothers of Cebu (Cebu Study Team, 1991, 1992; Guilkey *et al.*, 1989). While we have acknowledged the differences as important, and generally carried out stratified analyses to better understand the structure of relationships within urban and rural communities, we have not yet provided an adequate explanation of why such urban–rural differences exist. This paper is an attempt to explore urban–rural differences in greater detail by posing and answering four main questions:

1. How do health outcomes differ in urban versus rural environments of Metro Cebu?
2. How do community, household and individual characteristics differ in urban and rural environments? Do these characteristics cluster together in meaningful ways?
3. How are the distinguishing features of urban and rural environments related to health outcomes? Do health outcomes differ because of differences in the patterns and distribution of the underlying variables in urban and rural environments? Do outcomes differ because the underlying variables have differential effects on intermediate variables (such as infant feeding practices) in urban versus rural environments?
4. Is a simple urban–rural classification adequate for understanding health outcomes and their determinants?

Our objective here is not to develop a complete model of child health outcomes, but rather to focus on urban–rural differences and the underlying determinants of these differences. We hypothesise two dimensions to urban–rural differences: first, environments that people face are different; and second, the way in which people respond to the environment (i.e. make decisions) is different. Factors pertaining to the first dimension include prices, availability of health care, improved water supplies, transportation, roads and electricity, that is, the milieu of external constraints and opportunities that make up one's environment. These factors are generally not under the control of the individual and are referred to as exogenous factors.

The second dimension of urban–rural differences is behavioural: that is, even if faced with the same opportunities or constraints, people's choices and preferences (such as infant feeding practices) may differ depending on where they reside. This may reflect different value structures, which themselves may be the product of historical differences in urban and rural settings. We refer to these choices and behaviours as intermediate variables.

Our analyses are directed toward these dimensions of urban and rural differences. That is, we want to identify the relationship of exogenous factors and health outcomes, as well as the relationship of exogenous factors to intermediate factors that affect child health outcomes.

Study design and sample

The CLHNS was conducted in Metro Cebu, Philippines, from 1983 to 1986. The study area is one of about one million inhabitants, and includes Cebu city (the second largest city in the Philippines), surrounding peri-urban developments, small towns, and rural areas in the mountains or on offshore islands. The study is particularly useful for answering the questions posed above because of the wide range of ecological settings, areas of population density, level of urbanisation, and socioeconomic conditions, represented in the sample.

This is a prospective, community-based longitudinal study. By using a single stage cluster sampling technique, 33 barangays (or communities) were randomly selected from a set of 95 urban and 143 rural barangays. All pregnant women in these barangays were identified, and those who gave birth during a one year period (April 1, 1983 – March 31, 1984) were enrolled in the study. For this paper, we focus on 3080 single live births. The sample can be considered representative of all single live births in the Cebu area during a one year period. Surveys were conducted in the homes of respondents during the sixth to seventh months of pregnancy, at birth, then at bimonthly intervals for two years.

Extensive data were gathered to describe characteristics of households and individuals, including household size and composition, environmental sanitation, income, assets, and education. In addition, community-level surveys documented factors such as prices of essential goods and services, availability of health facilities, and community organisations. Thus there are ample data available to describe in detail the environmental circumstances in which children are born and raised. In addition, we have detailed time-varying information on health behaviours, feeding practices, use of preventive and curative health services, diarrhoeal and respiratory morbidity, and child weight and length.

For our analyses, we examine health outcomes at one year of age. At this point in time, the sample consists of 2506 infants, 75% residing in urban and 25% in rural communities. Loss to follow-up occurred primarily as a result of migration from the Cebu area ($n = 457$) and infant deaths ($n = 92$). Since we are examining environmental characteristics for individuals at the time of entry into the study (during pregnancy), we also excluded 25 cases where families had moved to different barangays within the study area by the time their children had reached one year of age. Since these exclusions may bias our parameter estimates, the results should be considered applicable only to non-migrants and infants who survive the first year of life.

We have chosen to focus on four health outcomes in infants: weight-for-age (WA), length-for-age (LA), weight-for-length (WL), and diarrhoea. Weight and length of infants were measured in the home by trained fieldworkers using Salter hanging-type scales and custom-designed length-measuring boards. We calculated Z scores of weight, length and weight-for-length using NCHS data. Diarrhoea and febrile respiratory infection data are based on the mother's report of her infant's symptoms during the seven days preceding each longitudinal survey. The number of surveys in which the infant was reported ill is used as an overall indicator of respiratory or diarrhoeal morbidity during the first year of life. We are also interested in feeding practices as important intermediate determinants of health outcomes. We used information on whether the infant was breast-fed on the day before each survey, and the level of supplementation with other foods. Caloric intake from these other foods was determined from 24 hour recall of all foods and liquids given to the infant on the day preceding each longitudinal survey.

Of the wide range of exogenous variables that describe the communities, households and individuals of the study, we have chosen to focus on a manageable set (of about 30) which we felt represented a number of significant categories of exposures. These include characteristics of the immediate environment (housing density, availability of safe water, excreta disposal practices, household hygiene); household income and assets; household composition; community-level prices of basic commodities such as oil, corn and kerosene; access to and distance to roads; availability of health services; maternal characteristics such as height, age, and education; sex of the infant; and season of measurement when the child was 12 months of age.

Initial classification of barangays as urban or rural was based on census track information, and basically indicates whether the barangay was contiguous with Cebu City. The sample includes 17 urban and 16 rural barangays. Urban barangays are characterised by high population density and are served by paved roads, public transportation, major medical centres, markets, department stores, etc. Rural communities are those separated from Cebu City by undeveloped land, farming areas, or water. Several of the rural communities are isolated in the mountains surrounding the city. Two communities are located on offshore islands and have no electricity or modern conveniences.

Results

Results and methods of analysis used to obtain them are presented to answer, in turn, each of the questions posed above.

Question 1: Do health outcomes differ in urban and rural areas of Cebu?

Figures 6.2–6.5 present mean values of Z scores for WA, LA, and WL, and diarrhoea prevalence for males and females in urban and rural environments at each bi-monthly survey during the first year of life. The most marked urban–rural differences are in morbidity, with urban children

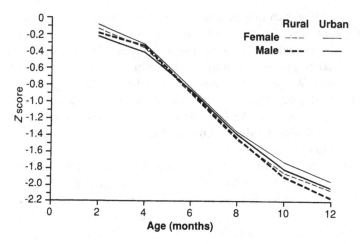

Fig. 6.2 Weight-for-age *Z* scores in Cebu infants aged 2–12 months, by urban and rural residence and sex.

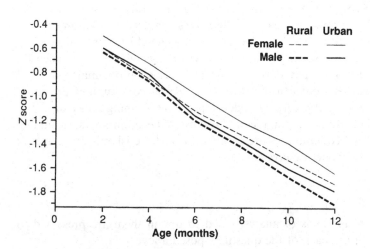

Fig. 6.3 Length-for-age *Z* scores in Cebu infants aged 2–12 months, by urban and rural residence and sex.

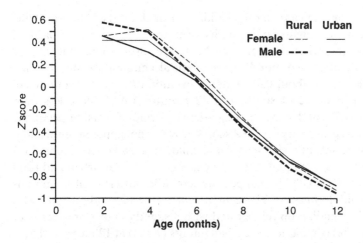

Fig. 6.4 Weight-for-length Z scores in Cebu infants aged 2–12 months, by urban and rural residence and sex.

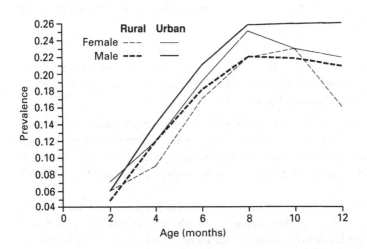

Fig. 6.5 Diarrhoea prevalence in Cebu infants aged 2–12 months, by urban and rural residence and sex.

having more diarrhoea than rural children throughout the first year of life. The pattern of weights and lengths over the first year is typical of developing countries: Z scores at birth are close to 0, but decline precipitously after two months of age. Rural children are shorter than urban children, by about 0.1 standard deviation (0.3 cm at 12 months of age). The prevalence of stunting at 12 months (LA Z score less than -2 s.d.) is 37.7% and 40.5% among urban and rural infants, respectively.

An urban–rural divergence in Z scores of WA is not apparent until about six months of age, at which point rural children show larger declines in Z scores than urban children. WL Z scores are positive for urban and rural children (with rural children appearing somewhat fatter) until six months of age. Rapid declines in WL begin at four months in both groups, with rural children ending up thinner and with a slightly higher prevalence of wasting (8.4%) compared with urban infants (7.5%) at 12 months of age.

There are also significant differences by sex of the child: boys consistently have lower Z scores of WA and LA, but the magnitude of the sex difference is greater in rural than in urban infants. The lowest prevalence of wasting at 12 months of age is found among rural females.

In summary, there are significant urban–rural differences in health outcomes among Cebu infants. While urban infants suffer higher morbidity, rural infants (particularly males) show a higher prevalence of linear growth retardation.

Question 2: How do community, household and individual characteristics differ in urban and rural environments? Do these variables cluster together in meaningful ways?

Mean values of the selected exogenous variables are presented in Table 6.1. At the community level, there are urban–rural differences in population density, housing density, price of kerosene and food items such as corn, oil and bananas, transportation access and roads, and availability of health services.

At the household level, more urban households have adequate excreta disposal facilities (e.g. dug latrines or flush toilets), electricity and safe water. Urban households have higher income and assets, including items such as television sets. They are closer to roads and have a shorter travel time to health facilities. Although the average size of urban and rural households is the same, urban dwellers are more likely to live in extended families, and have slightly fewer pre-school children. At the individual level, parental education is significantly higher in urban families, and urban mothers are more likely to have health insurance.

To see the extent to which the selected exogenous variables tend to

Table 6.1. *Mean values of exogenous variables in urban and rural communities*

	Rural (n = 633)		Urban (n = 1873)	
	mean	s.e.	mean	s.e.
Housing density[a]	2.01	0.03	3.32	0.02
Safe water available[b]	0.75	0.02	1.00	0.00
Poor excreta disposal[b]	0.68	0.02	0.25	0.01
Excreta around house[b]	0.35	0.02	0.33	0.01
Household assets (pesos)[c]	4671.71	412.83	11 886.7	899.65
Household income (pesos/week)	183.76	13.49	301.22	10.56
Own TV[b]	0.04	0.01	0.22	0.01
Own livestock[b]	0.82	0.02	0.51	0.01
Corn price	239.24	1.48	231.57	0.82
Kerosene price	471.53	5.93	493.99	3.85
Canned milk price	132.33	0.67	120.12	0.29
Formula price	378.78	3.08	359.75	1.58
Banana price	19.29	0.23	23.90	0.14
Oil price	99.96	1.23	99.14	0.76
Cerelac price	312.15	1.18	305.62	0.45
Household has electricity[b]	0.17	0.01	0.58	0.01
Access to roads[b]	0.26	0.02	1.00	0.00
Distance to road (m)	952.75	66.40	95.10	4.91
Mother's height (cm)	150.38	0.19	150.64	0.12
Mother's age (yrs)	26.90	0.25	26.11	0.14
Mother's education (yrs)	5.44	0.11	7.46	0.08
No. of years mother worked	4.25	0.21	4.16	0.11
Health insurance coverage[b]	0.04	0.01	0.12	0.01
Extended family residence[b]	0.27	0.02	0.42	0.01
No. of pre-school children	1.69	0.04	1.53	0.03
Household size	5.59	0.10	5.72	0.07
No. of persons per room	2.68	0.06	2.62	0.04
Health clinic hours open per week	58.68	1.69	126.88	1.35
Travel time to health clinic (min)	33.36	0.99	11.30	0.13
Sex of infant (proportion male)	0.51	0.02	0.53	0.01

[a] Relative scale 0–6, with 6 representing highest density.
[b] Dichotomous variables; value represents proportion with described characteristic.
[c] U.S. $1.00 = 20 pesos.

cluster together, we performed a principal component factor analysis on the full set of exogenous variables. We identified six factors, which together account for 47% of the total variance in the selected variables (see Table 6.2).

In factor 1 we find high loadings of housing density, availability of safe water, good excreta disposal practices (e.g. use of dug latrines or flush toilets), electricity, roads, hours that a health clinic is open, travel time to

Table 6.2. *Unrotated factor pattern: principal components factor method*

	Factor 1	Factor 2	Factor 3	Factor 4	Factor 5	Factor 6
Eigenvalue	4.93	2.32	2.25	1.75	1.55	1.47
Proportion of variance explained	0.16	0.08	0.07	0.06	0.05	0.05
Factor loadings	Housing density 0.62	Household income 0.52	Excreta around house 0.30	Prices: Kerosene 0.57	Years mother worked −0.46	Price of corn 0.50
	Safe water 0.55	Household assets 0.55	Children 0–6 0.63	Oil 0.60	Mother's age −0.52	
	Poor excreta disposal −0.64	Own TV 0.49	Household size 0.72	Cereal 0.53	Mother has health ins. −0.23	
	Electricity 0.62	Own livestock 0.40	Crowded household 0.73		Price of infant formula 0.43	
	Roads 0.63	Extended family 0.45				
	Distance to road 0.50					
	Mother's education 0.60					
	Father's education 0.64					
	Open clinic hours 0.52					
	Travel time to health facility 0.70					

health clinic and parental education. This factor appears to be a measure of what is usually thought of as urbanicity, or the extent of government infrastructure and commercial development in an area. Factor 2 represents household income, household assets (including ownership of radio, television, livestock) and an extended family residence pattern (which contributes to higher family income). Factor 3 is a household sanitation or crowding factor on which we find high loadings of household size, number of children 0–6 years of age, high number of persons per room and presence of excreta around the house. Factor 4 represents prices, and factor 5 represents a set of maternal characteristics (years the mother worked, age, mother has health insurance). Factor 6 has only one variable, the price of corn, with a high loading. Of all prices, corn shows the most variability in different settings.

Factor scores were computed for each individual. We then looked at the correlation of factor scores with urban residence. Factor 1 is highly correlated ($r = 0.69$) with urban residence. Factor 2 is significantly correlated with urban residence, but at a much lower level ($r = 0.25$). The other factors show little relationship to urban residence. Thus it is primarily those variables in factor 1 that together represent salient features of the urban environment.

In sum, the characteristics that vary most between urban and rural environments include those related to infrastructure (housing density, transportation access and roads, availability of health services) and sociodemographic factors (household sanitation, household composition, income and assets, and education).

Question 3: How are the distinguishing features of urban and rural environments related to health outcomes?

To answer this question, we first estimated a set of ordinary least squares (OLS) regressions in which the four health outcomes were dependent variables, and the 30 exogenous variables were the independent variables, and urban and rural samples were combined.

The exogenous variables have differential effects on the health outcomes. They also differ with respect to how much of the variability in outcomes they explain. The exogenous variables account for 17% and 20% of the variability in WA and LA, respectively, but only 5% of the variability in WL and diarrhoea. Several exogenous variables (e.g. excreta disposal facilities, distance to nearest road, health insurance coverage) had no significant effects on any of the health outcomes in the combined urban–rural sample.

Anthropometric outcomes (WA and LA) were most strongly influenced

Table 6.3. *Significant determinants of health outcomes in urban and rural infants*[a,b]

Health outcome	Weight-for-age		Length-for-age		Weight-for-length		Diarrhoea	
	urban	rural	urban	rural	urban	rural	urban	rural
Housing density	– –						+ + +	+ +
Excreta around the house			–				+ + +	
Household assets		+						
Income			+				+ +	
Electricity	+ +		+ + +		+ + +			
Own TV	+ +		+ + +		+ +			
Own livestock					+ +			–
Mother's education	+ + +		+ + +		+		–	
Extended family residence	+ +	+ +		+ +				
No. of children 0–6	– –		– – –					
Household size[c]	– –		– –	–	– –			
Crowding[d]	–		– –	– –				
Sex of the infant			+ + +				+ + +	
Mother's height	+ + +	+ + +	+ + +	+ + +				–

[a] Other variables included in the model but whose coefficients were not statistically significant for any of the health outcomes at $p < 0.10$: availability of safe water, type of excreta disposal facility, access to transportation, distance to the nearest road, number of years the mother worked, mother's age, mother's health insurance coverage, travel time to health facility.

[b] Symbols indicate the sign of the regression coefficient and level of significance (+, –, $p < 0.10$; + +, – –, $p < 0.05$; + + +, – – –, $p < 0.01$).

[c] Number of persons living in the household.

[d] Number of persons per room.

by height and education of the mother, sex of the infant, and variables representing household income, assets and household composition, with higher income, as expected, contributing to higher Z scores. Crowding within the house (high number of persons per room) decreased WA and LA, and large numbers of children 0–6 years of age also had significant negative effects on Z scores, whereas living in an extended family had a positive effect. High housing density, male sex of the infant and presence of excreta around the house contributed significantly to higher diarrhoea prevalence.

To test whether there were aspects of the urban environment unaccounted for by the selected exogenous variable, we ran the same sets of regressions described above, but added the variable indicating urban residence to the models. For WA and WL, the urban variable was not significant. However, for LA and diarrhoea, there was a significant additional effect of urban residence, in each case suggesting a detrimental effect of the urban environment not accounted for by the other exogenous variables in the model. This effect may reflect unmeasured exogenous exposures, or differences in intermediate factors or behaviours that are affected by the exogenous variables.

To see whether there was a different structure of relationships between the exogenous variables and health outcomes in urban and rural environments, we looked at the same regression models stratified by residence (see Table 6.3). In the rural strata, there were fewer significant effects to be found. For example, none of the selected exogenous variables were statistically significant predictors of WL in the rural sample. In general, the models identified factors that were statistically significant in the urban stratum, but not significant in the rural (although of the same sign). In a very few cases a variable was significant in the rural but not the urban strata (for example, there was a positive effect of household assets on WA, and a positive effect of extended family residence on LA in rural, but not urban, infants).

To further explore the differences in these models, we estimated the same models using a combined urban–rural sample with a full set of urban interaction terms. The significant interaction terms included mother's education for WA, WL, and diarrhoea, owning a TV for WA, and the number of pre-school children for LA. Since interaction terms have particularly low power (because of collinearity with their main effects) these differences are the most striking.

There are four possible explanations for the observed pattern of urban–rural differences. First, the rural sample is one third the size of the urban sample, so the power to detect a statistically significant result is greatly reduced. Second, the distribution of the exogenous variables is quite

Table 6.4. *Mother's education and infant feeding practices*

| | Education level | | | | | |
| | Primary | | Secondary | | Post-secondary | |
	urban (n = 753)	rural (451)	urban (892)	rural (136)	urban (169)	rural (12)
Percentage breast-fed						
2 months	0.91	0.97	0.79	0.88	0.56	0.67
4 months	0.88	0.96	0.71	0.84	0.46	0.58
6 months	0.85	0.95	0.65	0.79	0.38	0.58
8 months	0.81	0.93	0.61	0.75	0.33	0.58
10 months	0.76	0.90	0.58	0.71	0.27	0.58
12 months	0.69	0.83	0.50	0.65	0.22	0.50
Percentage exclusively breast-fed						
2 months	0.55	0.76	0.36	0.58	0.15	0.25
4 months	0.46	0.55	0.29	0.41	0.16	0.17
6 months	0.11	0.12	0.05	0.05	0.01	0

different for the urban and rural strata. This may lead to different parameter estimates if the true relationships are non-linear, but the parameters are estimated assuming (as we have) that a linear model is correct. Third, an exogenous variable may have different meanings in urban and rural environments. For example, the absence of dug latrines or flush toilets in rural areas represents less of a health hazard, because people in these areas use fields or places more distant from the house for defecation. The fourth explanation is that the exogenous variables differentially affect behaviour in urban and rural environments, possibly because of differences in the interaction of underlying variables. This explanation requires additional exploration.

Breast-feeding and caloric intake from supplemental foods are important intermediate factors that affect growth and morbidity. We looked at the effects of the set of exogenous variables described above on these variables by using linear regression models. Results show urban–rural differences in variables that influence feeding practices. Significant determinants of breast-feeding only in the urban sample include electricity in the household and number of years the mother has worked, both of which have negative effects on breast-feeding, and ownership of livestock and a higher number of pre-school children, which had positive effects. Ownership of a television, and health insurance coverage (both infrequent but indicative of higher SES), had negative effects on breast-feeding only in rural women.

The only shared determinant of feeding practices was mother's education. However, although more education decreases breast-feeding and increases caloric intake from supplemental foods in both urban and rural samples, the magnitudes of these effects differ. The effect of education on breast-feeding is nearly two times greater in the urban than in the rural sample. Another way to illustrate this finding is to look at breast-feeding patterns among urban and rural women with similar levels of education. At all levels of education rural mothers breast-feed their infants more: they are more likely to exclusively breast-feed infants up to six months of age, and to continue breast-feeding for longer periods of time (see Table 6.4). Thus, maternal education is likely to exert differential effects on health outcomes in urban and rural infants, in part because of its differential effect on infant feeding.

In sum, exogenous variables (especially those that reflect exposure to pathogens) have different direct effects on health outcomes in urban and rural environments. Moreover, exogenous variables also influence behaviours (such as breast-feeding) that are more proximate determinants of child health outcomes. The specific exogenous factors that are important, and the ways in which they ultimately affect health outcomes, differ in urban and rural environments.

Question 4: Is a simple urban–rural classification adequate for understanding health outcomes and their determinants?

Up to this point we have used conventional definitions of urban and rural to illustrate differences in health outcomes and their exogenous determinants. Urban–rural classifications have commonly been used by researchers as meaningful representations of factors which are assumed to occur jointly in one environment and not in the other. While the classification is a simple and useful method for explaining health outcomes, it obscures much of the heterogeneity that exists within urban and rural populations.

Gathering household-level data to describe all the components of the physical environment represents the other extreme. Although this approach would better represent the variation in the physical environment faced by each individual, it would not be an efficient, or in most cases even a feasible, method of data collection.

As an alternative between these extremes, an experienced fieldworker identified homogeneous areas within each barangay (the primary sampling units of the study), based on the type and density of housing, commercial and agricultural activities, provision of services and access to developed areas. Six settlement types were identified (the percentage of the sample

Table 6.5. *Health behaviours and outcomes by settlement type*

	Settlement type											
	Urban (n = 496)		Urban squatter (n = 361)		Peri-urban (n = 595)		Town (n = 470)		Rural (n = 219)		Isolated (n = 293)	
	mean	s.d.	mean	s.d.	mean	s.d.	mean	s.d.	mean	s.d.	mean	s.d.
Z score at 12 months												
Weight-for-age	-1.82	0.99	-2.02	1.03	-1.84	1.00	-1.88	1.01	-1.84	0.89	-1.99	0.85
Length-for-age	-1.64	1.03	-1.77	1.03	-1.73	1.01	-1.70	1.10	-1.99	1.02	-1.88	0.90
Weight-for-length	-0.86	0.89	-1.00	0.85	-0.80	0.91	-0.88	0.84	-0.89	0.75	-0.86	0.81
Diarrhoea[a] (0–6)	0.98	1.05	1.44	1.17	1.01	1.03	1.02	1.02	1.13	1.00	0.91	1.04
Number of surveys baby was reported breast-feeding (0–6)	3.95	2.42	3.87	2.46	4.35	2.28	4.56	2.19	4.55	2.25	5.55	1.27
Caloric intake of supplemental foods[b] (kcal)	2082	1607	2010	1562	1656	1439	1508	1344	1357	1155	683	888

[a] Sum of reported incidence in week prior to each survey (2, 4, 6, 8, 10, 12 months).
[b] Sum of intakes from six 24h recalls (2, 4, 6, 8, 10, 12 months).

represented by each type is indicated): urban (high population density, part of the two major cities; 19.8%), urban squatter (very high density, poorly constructed housing; 14.4%), peri-urban (lower density areas on the city borders, with good access to the urban centres; 23.7%), towns (small centres with governments and markets, but not contiguous with the major cities; 18.8%), rural (areas far from the urban centres, but with adequate transportation access; 11.6%), and isolated rural (poor access, sparsely populated; 11.7%). The largest proportion of respondents lived in peri-urban areas, followed by urban areas and towns.

Tables 6.5–6.7 present mean values of health outcomes, exogenous variables, and factor scores in each of the six settlement types described above (urban, urban squatter, peri-urban, town, rural, and isolated rural). Peri-urban settlements and towns share a number of exogenous character-istics, and are quite similar in terms of health outcomes. The two settlement types that clearly stand out as different from the others (primarily in illustrating the extremes in many of the characteristics) are the urban squatter and isolated rural.

Compared with all other types, urban squatter settlements are the most crowded: they have the highest population density, highest housing density, and highest level of crowding within the household. They have the poorest overall sanitation, with excreta frequently present outside the house. Safe water is, however, available in squatter areas from either deep boreholes with hand pumps (the most common source) or the municipal Cebu piped supply. Families in squatter settlements are poor; lower levels of household assets are found only among isolated rural families. Based on these observations, it is not surprising to find that health outcomes are poor in squatter settlements. Infants in these settlements have the lowest mean WA and WH Z scores, and the highest prevalence of diarrhoea. In addition, urban squatter infants are much less likely to be breast-fed than infants in any other settlement type except the non-squatter urban, where mothers are more affluent and educated. Thus, the group for whom breast-feeding is likely to be most important for preventing infection (that is, the group with the highest levels of diarrhoea) is the group least likely to be breast-fed (see Popkin *et al.*, 1990).

Isolated rural communities represented the other extreme in many community and household characteristics. They have the lowest popula-tion and household density, lowest household income and assets, poorest access to roads, transportation and safe water, longest travel time to health facilities, highest prices for foods (except bananas), lowest levels of maternal education, and highest number of pre-school children in the household. Levels of breast-feeding in rural isolated settlements are the highest in the sample: infants are breast-fed more frequently, supplemented

Table 6.6. *Mean values of exogenous variables by settlement type*

| | Settlement type | | | | | | | | | | | |
| | Urban | | Urban squatter | | Peri-urban | | Town | | rural | | Isolated rural | |
	mean	s.e.	mean	s.e.	mean	s.e.	mean	s.e.	mean	s.e.	mean	s.e.
Housing density[a]	3.64	0.03	4.23	0.03	2.92	0.03	2.91	0.03	2.35	0.04	1.26	0.03
Safe water available[b]	1.00	0.00	1.00	0.00	1.00	0.00	0.89	0.01	0.98	0.01	0.65	0.03
Poor excreta disposal[b]	0.07	0.01	0.19	0.02	0.28	0.02	0.37	0.02	0.65	0.03	0.87	0.02
Excreta around house[b]	0.31	0.02	0.61	0.03	0.31	0.02	0.23	0.02	0.16	0.02	0.41	0.03
Household assets (pesos)	11830.1	1471.4	8465.64	1148.3	9760.74	1112.3	13894.1	2741.5	9019.47	982.10	4555.08	628.38
Household income (pesos/week)	350.14	27.36	312.29	21.39	270.89	14.02	243.69	18.40	255.41	25.49	150.71	15.48
Own TV[b]	0.30	0.02	0.23	0.02	0.18	0.02	0.14	0.02	0.11	0.02	0.01	0.01
Own livestock[b]	0.43	0.02	0.41	0.03	0.56	0.02	0.63	0.02	0.77	0.02	0.92	0.02
Corn price	227.13	1.57	222.36	1.67	233.44	1.48	239.24	1.59	241.25	2.20	241.24	2.31
Kerosene price	497.94	8.97	470.42	11.76	525.73	3.96	410.34	5.03	507.29	7.85	524.38	9.54
Canned milk price	117.90	0.43	118.08	0.61	120.30	0.46	124.69	0.55	130.54	1.02	134.74	1.23
Formula price	367.33	3.30	351.40	3.52	361.76	2.63	360.59	3.02	371.52	4.55	381.16	4.83
Banana price	25.63	0.26	25.65	0.31	24.33	0.24	19.80	0.29	19.39	0.26	19.06	0.34
Oil price	90.37	1.20	97.51	1.68	101.61	1.44	99.63	1.51	113.43	2.07	97.80	1.64
Cerelac price	304.87	0.80	306.60	1.14	306.44	0.88	309.95	1.29	303.92	1.10	312.85	1.49
Household has electricity[b]	0.73	0.02	0.64	0.03	0.52	0.02	0.38	0.02	0.34	0.03	0.06	0.01
Access to roads[b]	1.00	0.00	1.00	0.00	1.00	0.00	0.60	0.02	0.79	0.02	0.24	0.03
Distance to road (m)	90.82	11.63	79.65	6.81	127.29	9.93	113.63	15.37	10.29	15.92	1864.12	120.81
Mother's height (cm)	150.90	0.23	149.75	0.26	150.80	0.20	150.86	0.22	150.65	0.28	150.03	0.29
Mother's age (yrs)	26.39	0.26	25.48	0.30	26.20	0.24	26.30	0.28	26.51	0.37	27.24	0.38
Mother's education (yrs)	8.49	0.15	7.50	0.17	6.83	0.13	6.81	0.14	6.37	0.18	4.69	0.16

No. of years mother worked	3.76	0.20	3.68	0.22	4.32	0.19	4.24	0.23	5.07	0.31	4.29	0.32
Health insurance coverage[b]	0.15	0.02	0.08	0.01	0.12	0.01	0.07	0.01	0.09	0.02	0.04	0.01
Extended family residence[b]	0.50	0.02	0.44	0.03	0.36	0.02	0.36	0.02	0.34	0.03	0.22	0.02
No. of pre-school children	1.51	0.05	1.53	0.06	1.53	0.05	1.59	0.05	1.66	0.07	1.70	0.06
Household size	6.07	0.14	5.80	0.16	5.43	0.11	5.68	0.12	5.69	0.16	5.42	0.14
No. of persons per room	2.82	0.08	3.06	0.10	2.47	0.06	2.51	0.07	2.35	0.09	2.61	0.09
Health clinic hours open per week	164.36	0.93	148.06	2.35	102.39	2.51	59.87	1.95	113.62	3.66	60.39	2.80
Travel time to health clinic (min)	8.98	0.11	10.28	0.14	13.51	0.28	14.99	0.45	15.62	0.77	49.44	1.51
Sex of infant (proportion male)	0.58	0.02	0.49	0.03	0.55	0.02	0.51	0.02	0.49	0.03	0.51	0.03

[a] Relative scale 0–6, with 6 representing highest density.
[b] Dichotomous variables; value represents proportion with described characteristic.

Table 6.7. *Mean factor scores by settlement type*

	Settlement type					
	Urban	Urban squatter	Peri-urban	Town	Rural	Isolated rural
Factor 1 Urban	0.77	0.62	0.15	−0.29	−0.36	−1.84
Factor 2 SES	−0.09	−0.46	−0.12	0.15	0.15	0.68
Factor 3 Household sanitation and crowding	0.05	0.28	−0.08	−0.001	−0.18	−0.12
Factor 4 Prices	0.09	0.18	0.15	−0.25	0.27	−0.63
Factor 5 Maternal characteristics	0.004	0.25	−0.05	0.02	−0.20	−0.09

less, and breast-fed for longer periods of time. They have the lowest levels of morbidity in the sample, but also show the highest prevalence of linear growth retardation, especially among males.

The limitation of the simple urban–rural classification is evident when we compare differences in mean health outcomes between settlement types with simple urban–rural differences. For example, the urban–rural difference in mean WA Z score amounts to 0.11 s.d., but the urban squatter–non-squatter difference is 0.20 s.d. Similarly, the difference in diarrhoea prevalence is much larger between urban squatter and non-squatter infants than between urban and rural infants.

These results demonstrate that a slightly more complex classification system can highlight important distinctions not apparent with the simple urban–rural dichotomy.

A further question is whether conditions in urban squatter settlements and isolated rural communities represent extremes of a continuum, or whether there are structural differences in the relationships between the underlying variables and the health outcomes in these areas. To explore this question, we estimated regression models of health outcomes in which the urban sample was stratified by residence in a squatter settlement. Despite the much smaller size of the 'squatter' stratum, a number of variables were significant predicators of health outcomes among squatters, but not in the rest of the urban population. For example, the presence of electricity in the household, and older maternal age, increased WA and LA Z scores and decreased diarrhoea; increased housing density increased diarrhoea, and excreta around the house decreased WL Z scores only in squatter infants. A surprising but very important finding is that mother's education, while highly important in the non-squatter sample, had no significant effect on health outcomes in squatter infants even among those whose levels of education were similar. There are at least two possible explanations for this finding. First, mothers in squatter settlements may be so constrained by other demands on their time and by lack of resources that, despite better education, they cannot adopt behaviours (such as hygienic food preparation) that promote health in their infants. Second, mothers in squatter settlements may have relatively little control over their infant's exposure to pathogens. The overall environment in squatter settlements is poor, and children share a great deal of potentially pathogenic outdoor space. Under these circumstances, even the more educated mother may not be able to protect her infant (especially since she also faces the constraints mentioned above).

In a parallel analysis, we looked at rural residents stratified by whether they were isolated or not. A different pattern of determinants of health outcomes emerged for isolated rural and non-isolated rural infants. Mother's height is the most powerful determinant of anthropometric outcomes in both rural strata. In isolated areas only, living in an extended family and ownership of livestock increase WA Z scores. Larger household size, longer travel time to a health facility, male sex of the infant, and having poor excreta disposal facilities decrease LA Z scores only in isolated areas. Mother's education is a marginally significant determinant of health outcomes only in isolated rural mothers (where education levels are generally the lowest in the entire sample).

In sum, the simple urban–rural classification scheme used by many investigators is useful, but limited. The most striking comparisons of health outcomes in infants emerged when a more refined definition of residence area was used. The distribution of important exogenous health determinants varies within the urban and rural strata. More importantly, the relationship between these factors and health behaviours varies as well, with similar levels of an exogenous variable producing different behaviours in urban and rural mothers.

Discussion

In the 1990s, it is insufficient to discuss urban–rural differences in child growth and health outcomes without acknowledging the heterogeneity of urban and rural environments. Special attention needs to be paid to two types of community: urban squatter settlements or slums, and isolated rural areas. In squatter or slum areas, there may be some advantages of urban living; for example, access to safe water, transportation, health care facilities, central markets, etc. At the same time, this part of the urban environment contributes to a higher risk of infectious disease.

What is pathogenic or unhealthy about the urban environment? There are several dimensions of pathogenicity that are apparent from our analyses. One is a reflection of the physical environment, and on the one hand includes environmental sanitation variables such as adequate excreta disposal facilities and cleanliness of the area surrounding the household, and on the other, factors that increase person to person contact and transmission of disease (population density, housing density, household crowding, number of pre-school children in the house).

A second dimension has to do with behaviours and choices that are promoted or constrained by factors operating within particular environmental settings. Thus, for example, we find that despite better education opportunities, women in urban squatter areas are less able to protect their infants from infectious disease than their counterparts living in rural areas. They are less likely to breast-feed, and this in turn may reflect the additional time and socioeconomic constraints faced by poor women.

It is particularly interesting to note, however, that despite the higher prevalence of infectious disease in urban (particularly urban squatter) settlements, it is rural children who experience a higher level of linear growth retardation. This reinforces the notion advanced by Waterlow (1988) and Keller (1988), that growth in length and in soft tissues are not affected by the same environmental variables. Chronically poor socioeconomic conditions (typical of isolated rural areas) are more often associated with stunting than with wasting. Keller (1988) cites cases where improvements in environmental sanitation (such as providing safe water), and reduced morbidity decrease the prevalence of wasting, but have little effect on stunting.

There are two possible explanations for the urban–rural differences in levels of linear growth retardation among Cebu infants. First, a genetic effect is possible. Rural Cebu mothers are, on average, shorter than urban mothers. A genetic explanation is not entirely justified, however, because we cannot be certain whether differences in maternal stature represent a genetic difference, or the mother's own nutritional history. A more likely explanation is that rural children may have inadequate diets. Chronic poverty and delayed supplementation with nutritionally inadequate weaning foods may lead to chronic energy deficiencies, or deficiencies of specific micronutrients such as zinc. This is something we can explore with the data at some point in the future.

The better linear growth performance of urban children in general, despite their higher prevalence of infectious disease, is also of interest. Despite higher morbidity risks in the urban environment, better nutrition or better health care may reduce the long-term effects of morbidity. However, in urban squatter areas, child health risks associated with early weaning and very high prevalence of infectious diseases are reflected in a high prevalence of wasting (10.6%) and a prevalence of stunting only slightly lower than that found in rural infants.

In sum, we have demonstrated marked urban–rural differences in (1) infant health outcomes that have a basis in differences in key underlying characteristics of urban and rural communities, most notably housing density, transportation access and roads, availability of health services, household sanitation, household composition, income and assets, and education; and (2) health behaviours that occur in response to the opportunities and constraints faced by individuals in urban and rural environment. Urban infants experience higher infectious disease morbidity and show a higher prevalence of wasting, but rural infants still exhibit more linear growth retardation. A simple urban–rural classification is useful for showing differences in health outcomes, but because of heterogeneity within urban and rural environments, a better classificatory scheme is one that highlights the special conditions that characterise urban squatter settlements or rural areas with limited access to the benefits of modernisation. This type of classificatory scheme allows us to see that the 'urban' pathogenicity effect is largely restricted to urban squatter areas.

Acknowledgements

This paper is based on research carried out during the Cebu Longitudinal Health and Nutrition Study, a collaborative research project involving the Office of Population Studies (OPS), University of San Carlos, Cebu, Philippines, directed by Dr Wilhelm Flieger; the Nutrition Center of the Philippines, directed by Dr Florentino S. Solon; and a group from the Carolina Population Center. Barry M. Popkin of the University of North Carolina at Chapel Hill is the project coordinator. Funding for parts of the project design, data collection, and computerisation was provided by the National Institutes of Health (Contract nos. R01-HD19983A, R01-HD23137 and R01-HD18880), the Nestlé Coordinating Center for Nutrition Research, Wyeth International, the Ford Foundation, the U.S. National Academy of Sciences, the Carolina Population Center, the U.S. Agency for International Development (AID) and the World Bank.

98   L. S. Adair et al.

References

Cebu Study Team (1991). Underlying and proximate determinants of child health: The Cebu Longitudinal health and nutrition study. *American Journal of Epidemiology* **133**, 185–201.
Cebu Study Team (1992). A child health production function estimated from longitudinal data. *Journal of Development Economics* **38**, 323–51.
Churchill, A. A. (1980). *Shelter.* Poverty and Basic Needs Series. Washington, D.C.: World Bank.
Guilkey, D. K., Popkin, B. M., Akin, J. S. *et al.* (1989). Prenatal care and pregnancy outcome in the Philippines. *Journal of Development Economics* **30**, 241–72.
Keller, W. (1988). The epidemiology of stunting. In *Linear Growth Retardation in Less Developed Countries* (Nestlé Nutrition Workshop Series, vol. 14, Nestlé Ltd) (ed. J. C. Waterlow), pp. 17–39. New York: Vevey/Raven Press.
Nutrition Foundation of India (1988). *Maternal nutrition, lactation and infant growth in urban slums.* New Delhi, Aga Khan Foundation Scientific Report No. 9.
Popkin, B. M., Adair, L. S., Akin, J. S., Black, R., Briscoe, J. & Flieger, D. (1990). Breast-feeding and diarrheal morbidity. *Pediatrics* **86**, 874–82.
Popkin, B. & Bisgrove, E. B. (1988). Urbanization and nutrition in low income countries. *Food and Nutrition Bulletin* **10**, 3–23.
Tanner, J. M. & Eveleth, P. B. (1976). *Worldwide Variation in Human Growth.* Cambridge University Press.
Underwood, P. & Margetts, B. (1987). Cultural change, growth and feeding of children in an isolated rural region of Yemen. *Social Science and Medicine* **25**, 1–7.
United Nations (1980). *Patterns of Urban and Rural Population Growth.* (Department of International Economics and Social Affairs. Population Studies No. 68) (ST/ESA/SER A68). New York: United Nations Press.
Waterlow, J. C. (ed.) (1988). *Linear Growth Retardation in Less Developed Countries* (Nestlé Nutrition Workshop Series, vol. 14, Nestlés Ltd). New York: Vevey/Raven Press.

7 Child health and growth in urban South Africa

N. CAMERON

The urban populations of the developing world are increasing at an unprecedented rate. In South Africa the black population has an urbanisation rate of 3.5%; this, coupled to a population growth rate of almost 3%, will lead to an increase in the urban black population from 6.5 million in 1985 to over 20 million by the year 2000. Such an urbanisation rate has major implications for a variety of support services, not the least of which is health care. This paper reviews current knowledge of the health status and growth of South African black children. Health status is examined through the broad measures of infant mortality rates, nutritional status, notifiable diseases, environmental factors and psychological measures of wellbeing. Physical growth is examined through a comparison of the latest studies describing the growth patterns of urban and rural children. Urban children of good socioeconomic status have growth patterns that are similar to those of the NCHS norms and significantly better than those of average urban children. Rural children from studies of farm labourers and subsistence farmers display significant differences in mean weights and body fat, but not height, and have superior growth to average urban children. The rationale, methods and pilot studies used in a new birth cohort study of children born in Johannesburg and Soweto are reviewed to highlight the methodological problems of urbanisation studies in developing countries.

Introduction

The urban environment of the Third World is increasing in size at an unprecedented rate. The latest WHO figures provide urbanisation rates of between 3.0% and 8.5% for Third World countries as defined by their high to very high level of mortality in the under-five age group (UNICEF, 1988). If this increase is added to the population growth rates for these countries then Third World urban environments will double, triple or quadruple in size over the next ten years.

First World or 'developed' countries commonly have urbanisation levels

99

greater than 75%. Thus the development of Third World countries will mean that in the future the majority of the population will live within urban environments. In 1958 Professor James Tanner described human biology as the study of 'nature in man and man in Nature' (Tanner, 1958). It is therefore wholly appropriate that human biologists should be turning their attention to the problems encountered by man in urban environments and indeed that the Society for the Study of Human Biology should choose urbanisation in the Third World as a topic for academic discussion.

The layman is often confused by terms such as the 'Third World' and 'developing countries'. These terms are often used synonymously by scientists and economists and yet the concepts of the 'Third World' and 'developing countries' stem from different approaches to the nations of the world. The term 'Third World' became common after the Second World War when Britain and her allies described themselves as the 'First World' and the communist bloc as the 'Second World'. The French demographer Alfred Suavy is credited with the first public use of the term 'Third World' in an article in *L'Observateur*, on 14 August 1952, entitled 'Trois mondes, une planete'. The Third World, he wrote, 'ignoré, exploité, meprise, comme le tiers état, veut lui aussi être quelque chose' (Worsley, 1984). Even though Suavy's use of the term may have been apt, the widespread acceptance of the term was not without its critics. Debray (1974), for example, described the term as 'a lumber room of a term, a shapeless bag in which we jumble together, to hasten on their disappearance, nations, classes, races, civilisations and continents as if we were afraid to name them individually and distinguish them one from another ... The term "Third World" indicates a certain backwardness in economic and social development [but] the real meaning of the "Third World" is that it presents the concept of a world apart, equidistant from the capitalist first world and the socialist second world, whose sole inner determining principle is that of underdevelopment...'. In practice the success of post-war economic and industrial rebuilding, predominantly in Europe and then in the Far East, created a functional or developmental distinction between these groups of nations such that the 'First' and 'Second' Worlds became the 'developed countries' and the 'Third World' became the 'developing countries'. This terminology is perhaps more appropriate because it implies a dynamic situation. If you are 'developing' you are in the process of change rather than being static, and that distinction is conceptually extremely important.

Those human biologists specialising in human growth and development appreciate that regular monitoring of growth status and growth rate are vital to the early identification and treatment of poor growth and development. The same is true of developing countries in that regular monitoring of man in these environments will allow early identification and

Table 7.1. *Urbanisation levels and urbanisation rates in South Africa, 1985*

	Urban (%)	Annual urbanisation rate (%)	Annual population growth rate (%)
White	89.6	—	1.6
Coloured/Asian	81.3	—	2.6
Black	39.6	3.5	2.9

treatment of poor development. Just as human variability means that each child is different, so too is each country; diagnosis and treatment must take account of these differences.

South Africa presents us with a unique case in point. There is little doubt that when one considers the sociopolitical, demographic, economic and health indicators of South Africa the 'Third World' or 'developing' status of the country is apparent. Yet South Africa also appears as a country caught between two worlds or states of change. Its Gross National Product of greater than U.S.$2000 per capita puts it within the top half of the world's nations, and yet its health indicators place it firmly in the bottom half. In common with oil-rich countries, the wealth of the country is controlled and owned by the privileged few, negating the health progress that could be achieved if wealth were more equally distributed. The underlying cause for this unequal distribution is, of course, the policy of separate development or 'apartheid', and it is important that as apartheid is cleared from the statute books so should human biologists be monitoring the inevitable social changes that will arise, so that they will be in a better position to diagnose and recommend treatment. Within this major social change the problem of urbanisation and its effects on child health and growth are of paramount importance.

Van der Merwe (1988), in an analysis of the latest (1985) census data, demonstrated that the white, Asian and so-called 'coloured' sections of the South African population had official urbanisation levels greater than 80% compared with 40% for South African blacks (Table 7.1). The use of the term 'official' is important because of the degree of unofficial or informal urbanisation that takes place in most Third World countries. In South Africa it is estimated that informal urbanisation accounts for a further 10%, making total black urbanisation just over 50%. The urbanisation rate for blacks was 3.5% per annum and the population growth rate 2.9% per annum in 1985. Van der Merwe (1988) estimates that these figures will lead to an increase of the urban black population from 6.5 million in 1985 to over 20 million by the year 2000.

Table 7.2. *Infant mortality rates in South Africa*

	National IMR	IMR in 10 urban areas	Johannesburg IMR
Whites	12.3	12.3	9.3
Coloureds	51.9	25.9	16.5
Asians	17.9	17.1	14.6
Blacks	85–124	38.6	27.0

Adapted from Yach (1988).

Rapid urbanisation has resulted in complex societal changes that have had both beneficial and adverse effects on the health of communities. The World Health Organisation (WHO) has recognised that the urban poor in developing countries are at greatest risk of several adverse health effects. Human biological and epidemiological research has a key role to play in planning to meet current and future health needs of urban communities by identifying those at greatest risk, identifying important risk factors that are amenable to intervention, and evaluating the effectiveness of intervention aimed at improving the health of urban communities (Yach *et al.*, 1990). The current health statutes of children in South Africa can be reviewed by using five broad measures of child health: infant mortality rates (IMR), nutritional status, notifiable diseases, environmental or ecological factors and psychological measures of wellbeing.

Table 7.2 illustrates a comparison of national, urban and Johannesburg–Soweto IMRs (Yach, 1988). Throughout these categories white IMRs are lower than other race groups and similar to those from the U.K., implying similar socioeconomic and health care conditions. Estimates for the black population are marred by under-reporting but illustrate clearly that all race groups in South Africa demonstrate improving IMRs as one moves into general urban areas and then the major urban conurbation of Johannesburg and Soweto. IMRs have decreased within race groups over the past few decades (Rip *et al.*, 1988; Bourne *et al.*, 1988; Herman & Wyndham, 1985) but the disparity between these groups has not significantly diminished (Yach *et al.*, 1991). Between 1929 and 1983, white IMRs decreased from 64.2 per 1000 live births to 13.5 per 1000. Coloured IMRs in the same period decreased from 158.5 per 1000 to 55.0 per 1000. Between 1970 and 1983 Herman & Wyndham (1985) demonstrate declines in IMRs for whites of 33%, coloureds 64% and blacks in Soweto 53%.

Information on growth and nutritional status has in the past been fragmentary, incomplete and, to a certain extent, outdated. Only in the past decade have a variety of studies been initiated to investigate the growth and nutritional status of urban and rural children; these, quite rightly, have concentrated on the black population who have been identified as being at greatest risk. Wagstaff *et al.* (1987) demonstrated that children attending 'average' schools in Soweto had mean heights and weights close to the NCHS fifth centile. Cameron (1992) showed that 'well-off' Soweto blacks had growth patterns similar to the NCHS norms and that rural blacks had better nutritional profiles than Wagstaff's average Soweto children.

The decline in childhood mortality and changes in recorded causes of death have been associated with declines in notifications for measles, polio, tetanus and, more recently, tuberculosis (TB) in children. The TB infection rate over the past 25 years has decreased by 5% in blacks, 7% in Asians and 8% in whites (Fourie, 1983). In spite of these declines, notifiable diseases in 1989 showed that infectious diseases such as measles, typhoid and TB constitute the bulk of annual notifications (Notifiable Diseases, 1990). Measles, polio and other EPI diseases (diseases within the Expanded Programme of Immunisation) would, of course, benefit from an expanded and integrated approach to immunisation. Studies of peri-urban and urban areas of South Africa indicate that recent migrants from rural areas have the lowest immunisation coverage (Yach *et al.*, 1991).

Von Schirnding (1988) has demonstrated that environmental pollution is already affecting the health of people living in urban and peri-urban areas of South Africa. Air pollution in particular constitutes a major health hazard and is thought to be mainly responsible for the increasing incidence of pneumonia as a leading cause of child mortality as opposed to a decreasing incidence of diarrhoeal diseases (Yach *et al.*, 1991). The air pollution problem is also, of course, associated with the relationship of smoking to health. Yach & Townshend (1988) and Yach (1989) have highlighted the impact of smoking on health and the rising smoking rate in children particularly in the urban and peri-urban areas. Long term consequences of exposure to such consumer products could result in dramatic increases in ischaemic heart disease and lung cancer in the next century.

Although it is recognised that comparisons of cognitive and social development across cultures and social classes is difficult because of the use of standard tests based on middle-class American samples, research in South Africa has demonstrated two trends: (1) the tested cognitive performance of black children has increased over the past decade; and (2) the psychological development of children during the first 18 months of life occurs at the same level and the same pace for all South African children (Richter *et al.*, 1989). After 18 months of age Richter & Griesel (1988) and Nerlove and Snipper (1981) showed that socioeconomic status and specific cultural experience begin to produce lower levels of performance for black children on standard tests compared with white children. Such observations are of considerable significance when compared with American results, which demonstrated more rapid development among American blacks compared with American whites. This suggests that, given very similar socioeconomic and cultural environments, black children develop more rapidly than white children.

This brief review of the health status of South African children, with a particular emphasis on the urban environment, demonstrates the dynamic relationship between health and urbanisation: infant mortality decreases, growth and nutritional status improves, and the incidence of notifiable diseases declines. The urban environment, however, also produces its own pattern of disease associated with the pollution endemic to rapidly expanding industrialisation. The effect of the urban environment on the cognitive and social functioning of the child in developing countries is not apparent in the infantile period, but its monitoring is hampered by inappropriate standard tests.

Growth pattern analysis

Human growth and development is recognised as the most sensitive indicator of child health and wellbeing; it is almost the exclusive domain of human biologists. Our research during the past few years has concentrated on describing the pattern of growth in rural and urban children through the medium of mixed-longitudinal studies.

Methods

Two rural areas were targetted for research. The first, called Ubombo, is in the northeast of South Africa close to the Mozambique border. Here the population relies on subsistence farming and financial support from the males of the family, who are mainly migrant labourers. Thus supplies of both food and money are seasonal and it is thought that migration to urban environments is a major social factor. The second, called Vaalwater, is a farming community to the northwest of South Africa. Here the population is stable. Housing, food and medical care are supplied by the farm owner, who is a paediatrician by training. There is little or no seasonal fluctuation in food or finance and the population is unlikely to migrate to urban environments.

Almost 400 children from both areas have been examined on a regular basis for the past five years. Anthropometric measurements of height, sitting height, weight, biacromial and bi-iliac diameters, head and arm circumferences and skinfolds at the tricep, bicep, subscapular and supra-iliac sites have been taken. In addition, assessments of secondary sexual development, using Tanner's (1962) staging technique, have been made at Vaalwater, and status quo studies of menarcheal age have been made at both venues (Cameron *et al.*, 1992).

Results

We have demonstrated that the children from these two different rural environments – farm labourers as opposed to traditional subsistence farmers – have similar growth patterns for height (Cameron, 1992) (Fig. 7.1). It is a pattern that we commonly associate with children living in the developing countries or Third World and is characterised by a gradual fall away from the NCHS fiftieth percentile between the ages of five and eleven years. By early adolescence almost half of the samples are below the tenth percentile, but adolescence is prolonged, creating the impression of a catch-up in growth such that adult heights are between the fiftieth and twenty-fifth percentiles.

There are, however, significant differences between mean weights for both males and females in these samples even though the pattern of growth is similar to that of height (Fig. 7.2). These differences are mostly due to

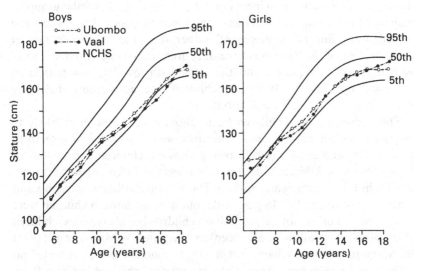

Fig. 7.1 Mean heights of rural black boys and girls from Ubombo and Vaalwater (Vaal) compared with the NCHS norms.

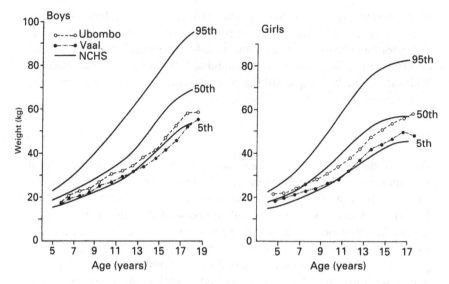

Fig. 7.2 Mean weights of rural black boys and girls from Ubombo and Vaalwater (Vaal) compared with the NCHS norms.

differences in subcutaneous fat thickness, implying, perhaps, either a genetic difference in the way in which fat is accumulated or differences in dietary intake favouring an increased fat content in the Ubombo group. It may also be the case that, as infants, the Ubombo group demonstrate the 'kwashiorkor line of development' rather than the 'marasmic line of development' of the Vaalwater farm-labourer group (Fig. 7.3): the former with high fat, low muscle and the latter with low fat, low muscle as suggested by Gurney (1969). As children and adolescents they may maintain this characteristic distribution.

The traditional rural children from Ubombo are those more likely to migrate to urban environments. Therefore we compared their growth to that of Wagstaff *et al.*'s (1987) average Soweto children and to that of well-off Soweto children. The latter group were children attending private schools in Johannesburg and Soweto. The well-off children had heights and weights close to the NCHS percentile, but average Soweto children were consistently inferior, not only to well-off children but also to rural children (Fig. 7.4). Such results have also been demonstrated by Malina *et al.* (1981) for Mexican children; Villarejos *et al.* (1971), amongst others, reported no differences between urban slum children and rural children in Costa Rica. It is well recognised that urban environments must be associated with improved socioeconomic status if they are to advantageously affect growth and health.

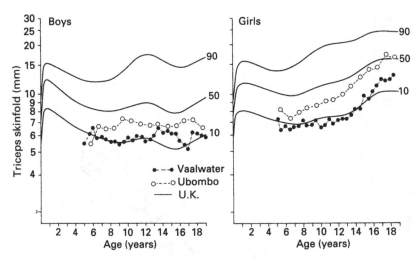

Fig. 7.3 Mean triceps skinfolds of rural black boys and girls from Ubombo and Vaalwater compared with the British (U.K.) norms.

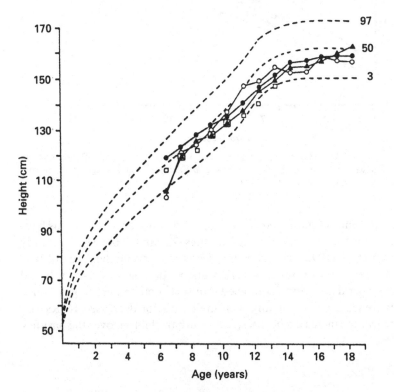

Fig. 7.4 Mean heights for black females from the sample of 'well-off' children from Soweto (open circles), rural children from Ubombo (filled circles) and Vaalwater (triangles) and 'average' children from Soweto (squares).

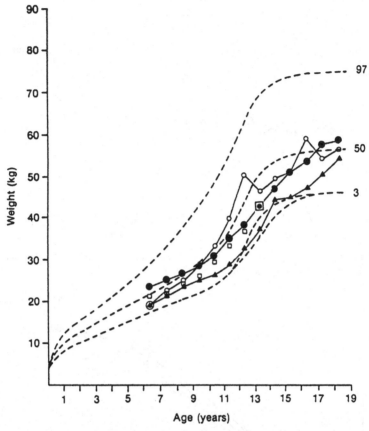

Fig. 7.5 Mean weights for black females from the sample of 'well-off' children from Soweto, rural children from Ubombo and Vaalwater and 'average' children from Soweto. (Symbols as for Fig. 7.4.)

Secular trends of improved growth and earlier maturational status have been demonstrated in South Africa (Figs 7.5 and 7.6). Most recently Cameron *et al.* (1992) have shown a decrease in menarcheal age of 0.34 years per decade for rural girls, 0.73 years per decade for urban girls and 0.46 years per decade for a combined group of rural and urban girls. Such secular trends are in sympathy with the trends for decreased IMRs and both reflect the increased health status of urban children over the past few decades.

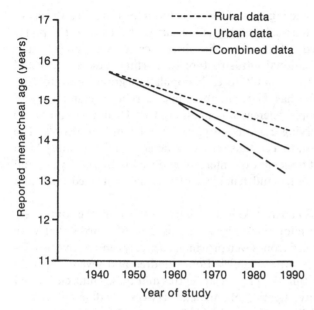

Fig. 7.6 Linear regressions of reported mean menarcheal ages against the year in which the study was undertaken. Data are from a variety of studies of South African black females from urban and rural areas reported by Cameron & Wright (1990).

Birth cohort studies: the 'Birth to Ten' project

Until 1988 we had concentrated our efforts on children over the age of five years, i.e. schoolchildren, because of the logistical difficulties of studying home-based children. Increasing concern about the effects of urbanisation on child health by the Medical Research Council led to the formation of a longitudinal birth cohort study of urban children. This study is entitled 'Birth To Ten' and has the major aim of determining the biological, environmental, economic and psychosocial factors that are associated with the survial, health, wellbeing, growth and development of children living in an urban environment (Yach *et al.*, 1990). The environment selected for this study is that of Johannesburg and Soweto. Soweto is situated about 20 km to the southwest of Johannesburg (hence SOuth-WEst-TOwnship). The apartheid policy of separate living areas for groups of different colours has resulted in distinct towns or 'townships' populated almost entirely by people of one skin colour. The gradual erosion of apartheid over the past few years has, however, created some non-racial areas, known locally as 'grey areas', in some of the more densely populated inner city areas in Johannesburg.

The organisation and logistics of Birth To Ten is the result of a variety of pilot studies that tested the feasibility of this birth cohort study. The policy of apartheid created a fragmented health service with large disparities between the organisational infrastructure supporting these services and between the facilities available to each population group. Soweto City Health, for example, has 11 well-baby clinics serving a community of almost 2 million people whereas Johannesburg City Health has almost 60 well-baby clinics serving a community of about 1.5 million people. The pilot studies for Birth to Ten thus set out to determine the monthly birth rate, the timing and frequency of antenatal and well-baby clinic visits and the availability, accuracy and reliability of routinely collected birth and growth data.

An average of 2680 births take place each month within the study area. The sample for the pilot studies was composed of 619 births that were proportionally sampled from seven provincial delivery centres, five provincial hospitals and six private clinics. Four hundred and seventy of these births took place within the provincial centres and these children formed the sample for the investigation of antenatal services and data on delivery and growth. The full sample of 619 births was used to investigate the feasibility of follow-up.

Eighty percent of the mothers received antenatal care and were first seen at a mean duration of pregnancy of 38 weeks. Women paid an average of five visits to the antenatal clinic and 89% of the women were seen at least twice prior to delivery. Of the 619 singleton births, 70 (11%) lived outside the study area and there were no traceable records of 37 (6%). A further 118 (19%) could not be immediately traced because the child was absent from the clinics (4%), they had moved away from the area (19%) or addresses were incorrect (10%). Sixty-seven percent of this sample could, however, be traced by actual visits to the houses.

Records of clinic attendance were available for 394 of the original sample of 619. These records showed that 350 (89%) of the children presented for all three of the DWT and polio vaccinations and thus were present at the clinics at six months of age. Three hundred and twenty-seven (83%) received a measles vaccination and thus were present at nine months of age.

Table 7.3 presents selected aspects of the data available from the delivery records. Virtually all mothers were checked for syphilis and for the presence of the rhesus factor. At admission blood pressure, fundal height and the presentation of the foetus were checked in 90% of the mothers, but the degree of cervical dilation was missing on the records of almost 24% of the sample. Apart from cord length, data relating to delivery were available on over 90% of the sample but data collected at the birth of the infant were less complete. Gestational age, the sex of the child and birth weight were

Table 7.3. *Delivery record information from Johannesburg–Soweto delivery centres*

Item	\multicolumn{4}{c}{Delivery record information}

Item	missing N	missing %	present N	present %
Syphilis serology	23	4.5	447	95.6
RH factor	14	3.0	456	97.0
Admission				
BP	53	11.3	427	88.7
Fundal height	56	11.9	414	88.1
Presentation	53	11.3	417	88.7
Cervical dilation	111	23.6	359	76.4
Delivery				
Presentation	44	9.4	426	90.6
Vaginal del.	15	3.2	455	96.8
Placental del.	41	8.7	429	91.3
Placental mass	37	7.9	433	92.1
Cord length	97	20.6	373	79.4
Infant				
Gestation age	12	2.5	458	97.5
Sex	4	0.9	466	99.1
Apgar 1	74	15.7	396	84.3
Apgar 5	144	30.6	326	69.4
Head circumference	80	20.0	390	80.0
Crown–heel length	82	17.5	388	82.5
Birth weight	2	0.4	468	99.6

available on almost the total sample, but Apgar scores and measurements of head circumference and crown–heel length were absent in about 20% of the sample.

There were major problems with both the availability of anthropometric equipment and measurement techniques at the clinics. Nine different scale types were being used, length was being measured either with a tape measure or a Pedobaby, and some clinics had no measuring equipment at all.

The results of these pilot studies indicated that the birth rate would support a potential sample of 3350 singleton births within a six week period. Twenty percent of these births would not fulfil residence criteria and a further 20% would be lost to follow-up because of untraceability or lost records. Over 80% of the original sample would be routinely available at nine months of age. Pregnant women could be contacted by 32 weeks of pregnancy to obtain permission for the study but this would limit antenatal information to the last trimester. Routinely collected growth data posed major problems and it was decided to supply purpose-built supine length measuring boards and tape measures to all clinics in addition to training the nursing staff in measurement techniques.

This series of pilot studies demonstrated that a birth cohort study in the Johannesburg–Soweto area was feasible. As a result the first mothers were enrolled in February this year and the first Birth To Ten baby was born on April 23 to a 40 year old mother of six children who presented at outpatients in the early stages of labour having had no antenatal care. This woman epitomises some of the problems that we hope that Birth To Ten will resolve; the problems of large families who live in overcrowded conditions and who should be the recipients of intensive health education programmes to cope with the problems of the urban environment. The logistical problems of such a study are exemplified by preliminary results relating to the sample size. Official (i.e. Governmental) notifications of births to mothers resident within the study area during the collection period amounted to 3846 out of a total of 4682; 836 births were to mothers resident outside the study area. Birth To Ten obtained delivery information on 2246 births and antenatal data on 2461 mothers; however, only 1566 of the mother–child pairs are within both delivery and antenatal samples. These statistics demonstrate the extreme mobility of people within our urban environments. The physical boundaries of the Johannesburg and Soweto Health Departments are entirely artificial divisions in the lives of people living and working on the Witwatersrand (the larger urban and industrial area within which Johannesburg and Soweto lie). Women living on the fringes of these urban health authorities move in and out of the areas, and thus the contact of the health services, depending on their preference and convenience. In short, the stability commonly associated with the urban environments of the developed world are simply not characteristic of our developing environment and thereby new research challenges are presented to the human biologist.

We believe that the importance of Birth To Ten, will, however, be profound not only in monitoring the effect of the urban environment on child health and growth but also in identifying its characteristics. Over 30 different research investigations will be carried out on the sample covering growth, psychological development, language and speech development, nutritional practices, dental health, morbidity, the effects of air pollution and the part that socioeconomic status plays in modifying the child's response to urbanisation. The findings will dictate health care policy and practice for the foreseeable future and will, we hope, result in healthier children in a post-apartheid, non-racial, democratic South Africa.

Acknowledgements

The research of the Human Growth Research Programme is financially supported by the Senate Research Committee of the University of the Witwatersrand, the Medical Research Council and the Foundation for Research Development. Birth To Ten is supported by the Medical Research Council, Anglo-American Chairman's Fund, Delmas Milling (Randfontein) Ltd and Kentucky Fried Chicken. These institutions and commercial companies are gratefully acknowledged. The research on urban and rural children would have been impossible without the assistance of Drs Farrant, Reid, Knight, MacDonald and Patel and Professor John Pettifor. Birth To Ten has been made possible through the collaboration of scientists and nursing staff from the Medical Research Council's Centre for Epidemiological Research in Southern Africa, the University of the Witwatersrand, the University of South Africa's Institute of Behavioural Sciences, Johannesburg City Health and Soweto City Health.

References

Bourne, D. E., Rip, M. R. & Woods, D. L. (1988). Characteristics of infant mortality in the Republic of South Africa 1929–1983. Part II: causes of death among white and coloured infants. *South Africa Medical Journal* **73**, 230–2.

Cameron, N. (1992). The monitoring of growth and nutritional status in South Africa. *American Journal of Human Biology* **4**, 233–4.

Cameron, N., Kgamphe, J. S., Leschner, K. F. & Farrant, P. J. (1992). Urban–rural differences in the growth of South African black children. *Annals of Human Biology* **19**, 23–33.

Cameron, N. & Wright, C. A. (1990). The initiation of breast development and age at menarche in South African rural and urban black females. *South African Medical Journal* **78**, 536–9.

Debray, R. (1974). *A Critique of Arms*. Vol. 1. London: Penguin Books.

Fourie, P. B. (1983). The prevalence and annual rate of tuberculous infection in South Africa. *Tubercle* **64**, 181–92.

Gurney, J. M. (1969). The arm circumference as a public health index of protein-calorie malnutrition of early childhood. *Journal of Tropical Pediatrics* **15**, 225–32.

Herman, A. A. B. & Wyndham, C. H. (1985). Changes in infant mortality rates among whites, coloureds and urban blacks in the Republic of South Africa over the period 1970–1983. *South African Medical Journal* **68**, 215–18.

Malina, R. M., Himes, J. H., Stepick, C. D., Lopez, F. G. & Buschang, P. H. (1981). Growth of rural and urban children in the Valley of Oaxaca, Mexico. *American Journal of Physical Anthropology* **54**, 327–36.

Nerlove, S. & Snipper, A. (1981). Cognitive consequences of cultural opportunity. In *Handbook of Cross-cultural Human Development*. (ed. R. H. Munroe, R. L. Munroe & B. B. Whiting), pp. 423–74. New York: Garland STPM Press.

114   *N. Cameron*

Notifiable Diseases, 1989. (1990). *Epidemiological Communications* **17**, 16–19.
Richter, L. M. & Griesel, R. D. (1988). *Bayley Scales of Infant Development: Norms for interpreting the performance of black South African infants.* Pretoria: Institute for Behavioural Sciences, University of South Africa.
Richter, L. M., Griesel, R. D. & Wortley, M. E. (1989). The Draw-a-man Test: a 50-year perspective on drawings done by black South African children. *South African Journal or Psychology* **19**, 1–5.
Rip, M. R., Bourne, D. E. & Woods, D. L. (1988). Characteristics of infant mortaility in the Republic of South Africa 1929–1983. Part I: Components of white and coloured infant mortaility rate. *South African Medical Journal* **73**, 227–9.
Tanner, J. M. (1958). The place of Human Biology in medical education. *Lancet* i, 1185–8.
Tanner, J. M. (1962), *Growth at Adolescence.* Oxford: Blackwell Scientific.
UNICEF (1988). *The State of the World's Children, 1988.* Oxford University Press.
Van der Merwe, T. J. (1988). A geographical profile of the South African population as a basis for epidemiological cancer research. *South African Medical Journal* **74**, 513–18.
Villarejos, V. M., Osborne, J. A., Payne, F. J. & Arguedes, J. A. (1971). Heights and weights of children in urban and rural Costa Rica. *Environmental and Child Health* **17**,m 31–43.
Von Schirnding, Y. E. R. (1988). Pollution – The growing threat to life. *Energos* **18**, 65–9.
Wagstaff, L., Reinach, S. G., Richardson, B. D., Mkhasibe, C. & De Vries, G. (1987). Anthropometrically determined nutritional status and the school performance of black urban primary schoolchildren. *Human Nutrition: Clinical Nutrition* **41C**, 277–86.
Worsley, P. (1984). *The Three Worlds.* Chicago: University of Chicago Press.
Yach, D. (1988). Infant mortality rates in urban areas of South Africa, 1981–1985. *South African Medical Journal* **73**, 232–4.
Yach, D. (1989). Urban marketing to promote disease or health? *Critical Health* **73**, 400–2.
Yach, D., Cameron, N., Padayachee, N., Wagstaff, L., Richter, L., Fonn, S., MacIntyre, J. & De Beer, M. (1991). Birth To Ten – Child health in South Africa in the 1990's. Rationale and methods of a birth cohort study. *Paediatric and Perinatal Epidemiology* **5**, 211–33.
Yach, D., Padayachee, G. N., Cameron, N., Wagstaff, L. & Richter, L. (1990). Birth To Ten – a study of children in the 1990's living in the Johannesburg–Soweto area. *South African Medical Journal* **77**, 325–6.
Yach, D. & Townshend, G. (1988). Smoking in South Africa. *South African Medical Journal* **73**, 391–9.

8 From countryside to town in Morocco: ecology, culture and public health

E. CROGNIER

Inroduction

In contrast with many other developing countries, urban life is traditional in Morocco. The main cities (Fes, Rabat, Marrakesh, Meknes) were built before the thirteenth century and have historically been, in a natural region or at the bounds of tribal areas, centres for trade, law, religion and learning, in agreement with the Muslim tradition. The secular interaction of these cities with their surrounding countryside remained balanced until the beginning of the twentieth century, through the domination of urban influence.

The modern type of urbanisation, aggregating a composite population, is also present. A first 'nucleus' including the harbour of Casablanca, Morocco's biggest town, and Rabat, the capital, absorbed most demographic increments until independence. From the 1960s onwards, however, these centres alone could no longer accommodate the rising velocity of population growth, and new urbanism spread to surrounding places, namely the harbours of Mahommedia, Sale and Kenitra. Thus, in contrast with tradition, modern urbanisation moved from inner provinces to the Atlantic shore, along with the translation of leading economic activity towards international trade.

In this process, the old cities (Fes, Marrakesh, Meknes), though also growing and maintaining their traditional functions, could not offer attractive prospects of employment. According to a Moroccan economist (Bentahar, 1987) they would become a relay between the small towns, the first levels of urban life, and the Atlantic conurbation, and instead of evolving toward increased urban patterns, would on the contrary be invested with a new 'ruralisation'.

What would be the consequence of this development? It would result in an absence of urban acculturation, a process that usually leads former peasants to lose not only their traditional occupations but also progressively their social, cultural and economic points of reference. Similarly, other

115

changes in basic behaviour patterns that urbanisation is supposed to entail would not occur, for example in those determining family size. These changes, their sequence and timing, are investigated in this work. The study is developed from observations collected in the province of Marrakesh and encompassing rural environments, small towns, and the city of Marrakesh itself.

The rural and urban environments in the province of Marrakesh

The province of Marrakesh is in inland South Morocco (Fig. 8.1). Its northern half is a partly irrigated agricultural plain; the southern half is mountainous, the ridge of the High Atlas Mountains rising very abruptly to summits of 3000 m or higher. A gross human partition is related to this division of the relief. The highlands are the estate of Berbers, while the villages in the plain are either Berber- or Arab-speaking, a blending that tends to be very intimate in urban centres.

Both geography and history predispose this region to agriculture and herding. Extensive farming is practised in the dry plain, herding and cultivated patches occur in the highlands, and large-scale agriculture takes place in the irrigated plain. There is consequently a wide span of rural ecosystems, from the early autarchic communities still practising a traditional life in remote valleys of the highlands, to the wide open areas assigned to an almost industrial management of citrus or olive fruits.

Local administrative services, schools, a rural hospital and shops around a market place are joined in four small towns (Amizmiz, Aït-Ourir, Imin-Tanout and Chichaoua), which make the first level of urbanisation. Their symbolic value as urban centres proceeds from material signs: density of human dwellings and the corollary development of services, electric light and running water in the houses. However, the rural lifestyle still dominates, since an important part of their population remains occupied in farming activities.

The city of Marrakesh is quite different. This old 'southern capital' is a melting-pot of northern and Saharan cultures. It is famous for its history, its architecture and its colourful population, a reputation which yearly attracts an important flow of international tourism. Its population exceeds 400 000 inhabitants (Ministère du Plan, 1982), either in the traditional medina or 'extra-muros', in recent and sometimes still urbanising extensions. In spite of the lack of industrial activities, the town is an attractive centre for the non-peasant population. Everyone can find electricity, running water, any kind of goods in a multitude of shops, a high human density, contacts with people coming from the whole world, and eventually employment, even if unskilled and insecure, in any activity except farming.

Fig. 8.1 The province of Marrakesh with its medical subdivisions and its urban centres.

Since 1982, several surveys have been undertaken in the Province of Marrakesh, to study the variations in female fertility and child mortality. They aimed at collecting a large data set about fertility trends and at observing their environmental correlates. The first survey concerned some 3000 inhabitants of the city of Marrakesh (Bley & Baali, 1987; Crognier, 1987; Crognier & Zarouf, 1987). A second province-wide survey in 1984 concerned the rural province and its small towns. One house in ten was visited and observations collected on some 5000 families (Bley & Baudot, 1988; Crognier, 1989). During the same period, local intensive investigations were also developed in several communities living in contrasted ecosystems: two villages in the irrigated plain (Baali, 1984), a set of peri-urban villages in the neighbourhood of Marrakesh (Hilali, 1985) and a series of villages in the highlands (Naber, 1989).

The total amount of information allows a tentative analysis of rural and urban polarities in relation to:

> data describing the conditions of life;
> data expressing several main cultural traits;
> data recording reproductive scores.

The shift in life conditions from rural to urban environments

In the study described here, life conditions were estimated by variables describing the socioeconomic frame and referring either to the dwelling:

> owner or tenant;
> conjugal or extended family dwelling;
> basic sanitary equipment (latrines) or not;
> lighting source: candle, petrol lamp, gas lamp, electricity;
> water supply from: river, well, spring, tank, running water;
> possession of radio or television set;
> means of transport: none, animal, bicycle, moped, car;

or to resources:

> number of wage earners in the household;
> number of persons in the household;
> profession of the woman questioned;
> profession of her husband.

A principal component analysis computed on the total sample of the provincial survey shows significant associations (Fig. 8.2): the rural–urban polarities are represented by the first factor, while the second shows contrasting levels of wellbeing. Some parameters, such as electricity, running water, lavatories, television set, car, moped or the absence of a vehicle, are an expression of urbanisation. Their typical owners are occupied in urban activities: employees or workers and to a minor extent shopkeepers or craftsmen. These parameters contrast with markers of countryside conditions in Morocco, such as water provided by a river, a well or a spring, more or less primitive lighting devices (candle, petrol lamp, gas lamp), the absence of lavatories, animal transportation, and indeed rural occupation: farmers and agricultural workers. Among the other items, some refer to economic conditions but do not discriminate between villagers and town dwellers: the number of rooms in the house, the number of persons in the household and the number of its wage earners. Several others have no particular significance such as the possession of a bicycle, which is very common, or the frequency of unemployment, equivalent in villages and towns. Some parameters are specific to the urban environment,

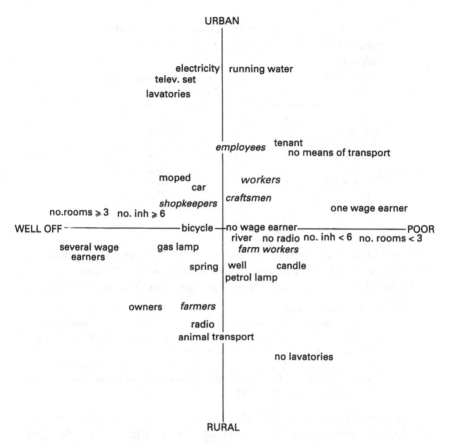

Fig. 8.2 The first two factors of a principal component analysis computed on the parameters estimating life conditions in the province-wide survey.

such as electricity and running water. Others, on the contrary, like the possession of a television set or a moped, are in no way linked to the urban condition. Their higher densities among town dwellers only express more typically urban needs.

On the basis of the presence or absence of at least one of the two: electricity or running water, a partition is made into urban and rural sub-samples. As the city of Marrakesh is not represented in the data, the urban sub-sample includes the four small towns of the province and therefore features the primary level of urbanisation.

This 'urban–rural' partition can also be considered as the alternative expression of a two-class variable, which would be made the dependent variable of a stepwise regression in which the other indicators of material conditions listed above would be the independent ones. The part of the

Table 8.1. *Stepwise regression of socioeconomic parametrers on the variable reflecting rural–urban polarities*

variables	step	partial R2	total R2
television	1	32.2	32.2
lavatories	2	4.1	36.3
no mean of trans.	3	3.7	40.0
owner/tenant	4	1.6	41.6
worker	5	1.0	42.6
several incomes	6	0.7	43.3
employees	7	0.5	43.8

variance explained by them would indirectly measure the variation in material conditions of life between urban and rural modalities.

Three methods have been tested to perform this stepwise regression:

1 a direct computation;
2 the computation of a principal component analysis, followed by the extraction of variables contributing most to each factor and their subsequent introduction in the regression;
3 the computation of a principal component analysis, followed by a varimax rotation and a subsequent stepwise regression taking the rotated factors as independent variables.

These procedures, tested in the different stages of this work, lead to similar results. In the present case, they converged to indicate that some 40% of the variance of the dependent variable could be accounted for by the indicators of life condition. Considering their different levels of sophistication, the best method appears to be the simplest, that is to say the closest to the raw information. For this reason only the results obtained through direct stepwise regression will be commented upon.

There is indeed a clear change in the material and economic environment when one leaves a *douar* (village) to live in a small city (Table 8.1). The greatest differences impinge first on the material environment; in successive ranks are the possession of a television set, of lavatories and of a means of transport. The differences in professional categories do not segregate small towns from villages, hence their minor contribution to the equation: the first to be included, at the fifth rank, is the category of workers, which accounts for only 1% of the variance.

This may signify that primary urbanisation does not induce sociological changes; however, more simply, it may also show that material markets are a better sign of urbanisation than is occupational organisation.

Table 8.2. *Comparison of several socioeconomic parameters in environments of increasing level of urbanisation*

characteristic (%)	rural province	small towns	Marrakesh
economic standard			
lavatories	23	73	88
TV set	4	56	71
owners	90	75	42
number of rooms	3.3	3.3	2.3
mean no. inhab./household	6.8	6.6	6.8
mean no. inhab./room	2.1	2.0	2.9
occupation of husband			
farmer	48	18	01
worker	14	35	50
craftsman	3	5	10
shopkeeper	8	14	16
employee	1	13	12
occupation of wife			
housewife	95	97	95
worker/craftswoman	4	2	4
employee	0	0	1

The comparison of data when the city of Marrakesh is included (Table 8.2) shows a steady trend toward the improved expression of urban traits: the level of equipment of the household improves regularly from countryside to city (for example, more lavatories and television sets) while on the contrary, the proportion of owners and the dwelling surface both regress. With household members remaining as numerous as in rural areas, this leads to the well-known effect of crowding in urban homes, a circumstance common to large cities, and appearing in Marrakesh while it is still absent from small towns.

The increase of urbanisation also entails changes in the distribution of professions. The obvious decline of farming activities is balanced by a threefold rise of secondary occupations (workers and craftsmen). The intrinsic urban employment represented by tertiary occupations is in the same proportion in either large or small towns.

A point to underline is the stability of females' social condition in either environment. Female employment remains uncommon, even in the city of Marrakesh, and is bound either to the lowest social categories or to the highest ones, or to broken matrimonial status. This low level of female employment is indeed related to females' status in Moroccan society and is certainly a part of Moroccan culture resistant to ongoing westernisation.

Table 8.3. *Stepwise regression of several culturally determined traits on the rural–urban variable*

variables	step	partial R2	total R2
husb. prim–sec educ.	1	7.9	7.9
wife without educ.	2	3.3	11.2
monogynous marriage	3	1.8	13.0
husb. koranic school	4	1.1	14.1
endogamous marriage	5	0.3	14.4
inbred mates	6	0.3	14.7
wife koranic school	7	0.2	14.9
divorced mates	8	0.1	15.0

The shift in sociocultural behaviours

The following information has been collected in relation to the provincial survey:

> matrimonial status: married, widowed, divorced;
> mates inbred or not;
> age of the woman at her first marriage;
> number of successive marriages;
> educational level of the husband: illiterate, koranic school, primary school, secondary school, other;
> education level of the wife: illiterate, koranic school, primary school, secondary school, other.

Their introduction as independent variables into the computation of a stepwise regression on the variable encompassing the rural and urban polarities, previously defined, shows no evidence of a change in cultural behaviours from village to small town (Table 8.3).

The predictive value of the equation is very poor, as only 15% of the variance is explained by the regression. Furthermore, it should be noted that most of the variables are assumed from information about the educational level of mates (12.5%), variables tightly associated with the pattern of occupations and with the distribution of schools in the area and therefore dependent not only on cultural patterns but also on economic structures. In particular, basic social behaviours related to marriage provide no evidence of contrasting ideas between countryside and small towns.

More information can be brought to bear on this point by the comparison of several population samples in the province of Marrakesh. Table 8.4 compares data collected in several Berber douars from a valley in

Table 8.4. *Comparison of matrimonial behaviours and educational levels between samples of rural and urban populations from the province of Marrakesh*

characteristic	Azgour	Aït Imm.	r. prov.	towns	Marrak.
matrimonial behaviour					
monogyny	90%	99%	90%	78%	86%
endogamous marriage	58%	59%	49%	53%	—
inbred mates	30%	10%	24%	20%	—
age at first marriage (years)	17	16	18	17	18
married only once	96%	83%	90%	86%	—
education					
wife no education	100%	100%	98%	85%	60%
wife koranic school	0%	0%	1%	3%	17%
wife public school	0%	0%	1%	12%	23%
husband no education	46%	—	74%	45%	25%
husband koranic school	37%	—	19%	25%	36%
husband prim./second. school	16%	—	7%	30%	49%

the High Atlas (Azgour) and in two Berber villages from the plain (Aït Immour) with the mean results obtained in the provincial survey for the rural part of the study, those obtained for the four small towns, and those of the city of Marrakesh. The variations in matrimonial behaviour from sample to sample are rather moderate when the divergence related to local situations is taken into account (e.g. the rates of inbreeding in local populations or the lower incidence of monogamy in small towns). In particular, the age of women at first marriage is stable; this may express its rigorous determination by social consensus in a still pre-Malthusian society.

As far as schooling is considered, the decisive importance of facilities again appears. No girls living in villages in the 1950s (the women questioned average 35 years in 1984) could get access to the few existing schools, not even to the far more frequent koranic schools, and neither could a majority of boys. The urban environment breaks this situation, first for boys and secondarily for girls. But if the frequency of schooling begins to increase in small towns, it only reaches an appreciable level in the dense urban environment of Marrakesh.

The shift in reproductive scores

Three categories of data are considered: observations related to reproductive behaviours, measures of fertility and of child mortality, and observations estimating the degree of medical care.

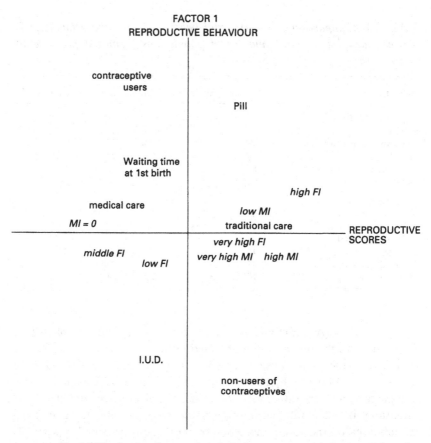

Fig. 8.3 The first two factors of a principal component analysis computed on the parameters related to reproductive scores in the province-wide survey.

Family planning:

use of contraceptives: never; in the past; now; in the past and now;
contraceptive method employed: pill; I.U.D.; condom; others;
waiting time at first birth.

Fertility and child mortality scores (computed according to Crognier, 1989):

mother's individual fertility index (FI): total no. of births / no. of births in age class;
mother's index of child mortality (MI): mortality rate in her progeny / raw mortality rate in her age class.

Health care:

therapeutic uses: traditional medicine; district nurse; medical centre; physician; traditional medicine and public health care;
number of visits to a medical centre during the last year.

Table 8.5. *Stepwise regression of reproductive parameters on rural–urban polarities*

variables entered	removed	step	partial R2	total R2
intrauterine device	—	1	0.020	0.020
district nurse	—	2	0.015	0.035
traditional medicine	—	3	0.010	0.045
permanent fam. plan.	—	4	0.003	0.048
low fecundity index	—	5	0.003	0.051
actual fam. plan.	—	6	0.001	0.052
never fam. plan	—	7	0.002	0.054
—	permanent fam. plan.	6	0.000	0.054
high mortality index	—	7	0.001	0.055
fecundity index $> x$	—	8	0.001	0.056
very high mortal. index	—	9	0.001	0.057

A principal component analysis may help us to understand in which terms their combinations result in differing reproductive scores (Fig. 8.3).

The first factor essentially summarizes the behaviours affecting reproduction (contraception, medical care), while the second expresses the differences in reproductive success. Several classical results appear: a strong association between the highest fertility and mortality indices, the absence of contraception and the use of traditional medicine (all markers of pre-Malthusian families) as opposed to the contraceptive users, whose habits are associated with medical care, with a delayed first birth and even with the absence of mortality in their progeny.

Two results are specially interesting in the Moroccan context: those of the IUD users and those of pill users. The first signals low fertility scores while the second is positively associated with rather high ones. This is because the pill users are frequently recent recruits to family planning, which they adopted with the idea of stopping the extension of an already large family, while the IUD users are mainly older recruits, often coming to the IUD after having been pill users. The stepwise regression of these variables, still on the same two classes of variable established from the province-wide survey and encompassing the primary urban and the rural polarities, fails to associate their variation with these differing environments, although the first two parameters introduced in the equation are on one hand the IUD, which necessitates medical care only accessible in towns, and on the other, the medical aid provided by the district nurses who work only in the country (Table 8.5).

The comparison of the mean results given by these same variables, not only in the rural province and in small towns, but also in the city of Marrakesh, yields differing results (Table 8.6). Although the differences are

Table 8.6. *Comparison of reproductive scores and of behaviours related to medicine in three environments of increasing urbanisation*

characteristic	r. province	small towns	Marrakesh
never use of contracep.	49%	44%	48%
contraception by:			
pill	93%	76%	78%
IUD	5%	10%	14%
health care by:			
tradit. medicine	14%	2%	1%
hospital dispensaries	48%	55%	67%
mixed tradit. & disp.	21%	26%	16%
no. of visits to hosp./year	3.0	7.1	12?
fertility index:			
childless	6%	6%	6%
low	7%	10%	12%
middle	33%	34%	38%
high	37%	34%	31%
very high	17%	16%	13%
mortality index:			
null	38%	43%	63%
low	20%	20%	6%
high	34%	30%	20%
very high	8%	7%	11%

effectively slight between the rural province and small towns, they become more important when the city of Marrakesh is considered. In particular, the data related to the number of progeny differ significantly between the three environments, because there is a relative increase of the frequency of families with a small number of children in the city of Marrakesh, and because there is a concomitant decrease in mortality scores. At a general level, the results are clear. The curves of cumulative fertility (Crognier, 1989) indicate that the mean number of live births for women aged between 45 and 49 years is 8.0 in the rural province but only 6.3 in Marrakesh.

A trend toward a more frequent use of public health services is indicated; it is probably for this reason that the main appreciable difference is observed in mortality indices, essentially in the clearly increased frequency of families with no child mortality, in Marrakesh compared with the countryside. A good illustration of this difference is given by statistics comparing death rates during the first three months of life, in a population of villagers in the neighbourhood of Marrakesh, still involved in a peasant lifestyle but having easy access to dispensaries (Hilali, 1985). They show a threefold reduction of mortality rates among babies vaccinated.

Table 8.7. *Changes in life conditions, cultural and reproductive behaviours, from countryside to town and city, in the province of Marrakesh*

Change	Do not change
life conditions	
freq. of owners	no. of inhab. in household
equipment of household	women's employment
professional categories	
dwelling surface	
sociocultural traits	
boys' schooling rate	age at first marriage
girls' schooling rate	matrimonial types
health and reproductive behaviours	
increase in medical aid	freq. pre-Malthus. fam.
drop of child mortality	freq. childless fam.
slight decrease in natality	

Conclusion

The overall results show that urban adaptation in Morocco is, as elsewhere, an evolving patchwork in which fast-changing behaviours are both the most dependent upon the man-made environment and the most labile, for example household equipment and professional categories (Table 8.7). When the environmental context and deep cultural options are simultaneously involved, as is the case regarding girls' schooling, family size or health care, changes are happening, though supposedly far from being completed. On the other hand, tradition essentially persists when cultural references are the leading parameters, as for example in the age at marriage, or in a wife's occupation.

The lack of comparative data concerning the developing conurbation denies us the appreciation of possible differences between Morocco's new urban life and life in provincial capitals. Furthermore, cross-sectional data are not suitable for the analysis of behavioural change through time. It is therefore not possible to know whether the provincial capitals are getting 'ruralised' or if they are still urbanising. Although the more suitable longitudinal approach to the question of changes induced by urban life is still to be taken in Morocco, the available data allow brief insights into the time dimension of urban acculturation. An example is given in Table 8.8, which considers changes in the age at first marriage over a twenty year interval and in different environments; as can be seen, there is no sign of a quick change in this cultural trait, in spite of the fact that Moroccans themselves agree in their estimate that girls get married later and later in

128   *E. Crognier*

Table 8.8. *Variations in mean age (years) at first marriage with respect to rural or urban dwelling, over an interval of twenty years*

age class (years)	Azgour	r. prov.	s. towns	Marrak.
45–49	17.11	17.82	17.08	18.05
	±3.25	±3.46	±2.82	±4.22
25–29	17.67	18.01	18.10	18.04
	±2.60	±2.31	±3.09	±3.51

cities, and that the age at marriage in town has always been later than in villages. In the same way that gene pools are slow to respond to environmental pressures, and thus balance the fast acclimatisation of organisms, is not the inertia of culture perhaps the best guarantee of men against their ever-changing minds?

References

Baali, A. (1984). *Les agriculteurs Ait Immour et leurs migrants en ville. Etude comparative des paramètres socio-économiques, démographiques, bio-médicaux.* Thèse de Doctorat de 3ème cycle, Fac. Sc. Univ. Cadi Ayyad, Marrakech, 160 pp.
Bentahar, M. (1987). *Villes et Campagnes au Maroc. Les Problèmes Sociaux de l'Urbanisation.* Rabat: Editell.
Bley, D. & Baali, A. (1987). La mortalité infantile et juvénile, analyse différentielle en fonction de l'environnement socio-économique. *Revue de la Faculté des Sciences de Marrakech* 3, 63–88.
Bley, D. & Baudot, P. (1988). Some recent trends in infant mortality in the province of Marrakech, Morocco: a demographic transition in process. *Social Biology* 33, 322–5.
Crognier, E. (1987). Child mortality and society in Morocco. *Journal of Biosocial Science* 19, 127–37.
Crognier, E. (1989). La fécondité dans la province de Marrakech (Maroc): enquête anthropologique. *Bull. soc. Roy. Belge d'Anth. et Prehist.* 100, 113–22.
Crognier, E. & Zarouf, M. (1987). Fécondité, mortalité et milieu socioéconomique dans la ville de Marrakech. *Revue de la Faculté des Sciences de Marrakech* 3, 7–62.
Hilali, M. K. (1985). *Etude bio-démographique et sanitaire des populations péri-urbaines de la zone d'El Azzouzia (province de Marrakech, Maroc).* Thèse de Doctorat de 3ème cycle, Fac. Sc. Univ. Cadi Ayyad, Marrakech, 151 pp.
Ministère du Plan (1982). *Population légale du Maroc d'après le recensement général de la population et de l'habitat de 1982.* Rabat; Direction de la Statistique. 215 pp.
Naber, N. (1989). *Etude de comportement fécond d'une population féminine de la haute vallée d'Azgour (cercle d'Amizmiz, province de Marrakech).* Thèse de Doctorat de 3ème cycle, Fac. Sc. Univ. Cadi Ayyad, Marrakech, 120 pp.

9 *Urban–rural population research: a town like Alice*

This paper describes the multiple research methods that can be utilised by researchers to describe and analyse complicated and dynamic processes which occur in populations. Because of the inherent limitations in our understanding of an event or process at a given point in time, we often need to draw on data obtained through different methods and from different sources if we are to obtain a composite picture and realistic analysis of this process. We therefore have to accept that the resulting analysis will itself need to be reviewed once new data have been obtained, because the data obtained by the various methods will have different levels of certainty associated with them. Each method has its own methodological problems, but when used in conjunction with each other, their consistency and validity can be evaluated.

Initially, a theoretical framework in which population-based research needs to be performed and interpreted is presented. This is followed by a case study, which will exemplify the use of these complementary research methods. A variety of methods were used to investigate the biological and cultural factors influencing the morbidity patterns of Central Australian Aboriginal children living in the Alice Springs' town-camps. Some concluding remarks will integrate the two sections within the context of population-based research. For the purpose of this discussion, the antithesis of population-based research is laboratory-based research, although the laboratory setting itself can constitute a population in its own right.

The model

It has been appreciated for a long time that notions like 'health' and 'disease' need to be interpreted in the cultural context and epoch from which they arise, because they are primarily socially constructed concepts (Kelman, 1975). This was recognised long ago by Greek authors, including Plato (Moravcsik, 1976), reiterated over the centuries and again recently by Blaxter (1990, pp. 13–34).

129

To assess an individual's state of health we therefore often have to rely on proxy indicators, which we assume reflect this status. For instance, height-for-age and weight-for-age in children have for a long time been considered to be indices of a child's nutritional status (Waterloo, 1976, pp. 535–7). Height-for-age is an index of chronic nutrition and weight-for-age an index of current nutrition. Whereas Waterloo (1976, p. 539) considers weight-for-height to be a more refined index of current nutrition and especially current energy intake, other authors prefer weight-for-(height)^2 (Babu and Chuttani, 1979).

Most authors recognise that the expression of health and disease is a complicated process, dependent on both cultural and biological factors. Again, child growth provides us with a prime example of this process (Cheek, 1968; Tanner, 1976). While the child's genetic endowment delineates that individual's biological potential, its expression is modulated and limited by the environmental conditions in which that child is reared. These environmental conditions include both the physical and the cultural factors operative in this context. The physical factors include the climatic and geological characteristics of the region as well as the housing conditions in which the child lives, including the prevailing sanitary conditions. Similarly, the cultural factors consist of the social and economic conditions that prevail within the group of which the child is a member. These result in the existence of certain behavioural patterns and norms which operate at a group level, some of which are adopted by the individual and are influential in shaping individual behavioural patterns. This highlights another relationship which exists between individuals and populations: while genetic factors may be found to be more influential at an individual level, at a population level the environmental factors are often found to be more important (Rose, 1985). Our conceptual framework needs to be able to accommodate these tensions, between biological and cultural influences, genetic and environmental factors, operating at both individual and population levels.

The conceptual framework that performs this task and can accommodate these different interactions is referred to as the Ecological Model of Disease Causation (Fig. 9.1). In this context, any observed phenomenon (or phenotype) results from a complex interaction between causative agent, host and environmental factors. While initially promoted by epidemiologists dealing with infectious disease problems (Barker, 1975), this framework can be extended to analyse the occurrence of any biological or illness phenomenon (Beck, 1985; Beck *et al.*, 1989). In this framework the agent is the final common pathway through which the relevant pathophysiological mechanisms are mediated. This includes microbiological agents (such as bacteria or viruses), carcinogens, excess or deficiency of hormones, caloric

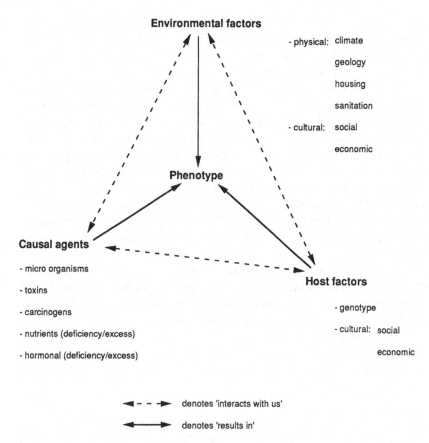

Fig. 9.1 The Ecological Model of Disease Causation, showing the interaction between agent, host and environmental factors.

or protein intake, or vitamins and other agents that interfere with the host's cellular metabolism. Agents may act in their own right or can be aided by coexistent factors (cofactors), which act synergistically with them. Host factors relate to the individual in which the process is occurring, for instance the stunted child. These factors include both the host's genetic endowment, which determines the development of the host's organ systems, and the cultural factors, which relate directly to that individual and influence his or her behavioural patterns. These cultural factors are subdivided into social and economic factors operating within that individual's domain. Both agent and host factors operate within a wider environmental context. These environmental factors, as described above, consist of the physical environment in which both host and agent exist and the cultural context to which they relate. The cultural context comprises the social and economic conditions that determine social mores and values at a group level and influence group behavioural patterns, which in turn influence individual behaviour and value systems.

This model integrates many of the tensions previously referred to into a comprehensive framework, which allows for the development of an integrated analysis of disease causation at both an individual and population level, while maintaining the cultural context in which this occurs. As such, the model is an extension of both the aetiological and social conceptions of disease causation (Riese, 1953). It also puts intractable debates, like the behaviouralist versus structuralist debate on disease causation (Townsend & Davidson, 1986) and the 'nature' versus 'nurture' debate, into a different perspective. While recognising the importance of either opposite, the specific importance of each, however, changes with different contexts and at different points in time.

The interdisciplinary nature of this interpretation of disease causation becomes apparent, drawing extensively on biomedical, biological and social sciences (Fig. 9.2). Those interested in population-based research need to be acquainted with all three aspects of this triad. Only then can we achieve more profound insights into the reasons for the occurrence of certain morbidity and mortality patterns in human populations. That this is particularly relevant to epidemiologists was recently reiterated by Susser (1989).

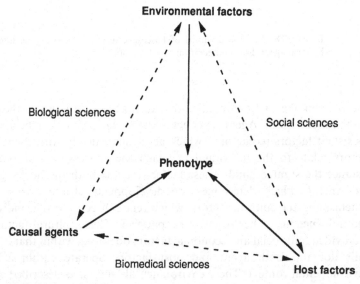

Fig. 9.2 The multidisciplinary framework of the Ecological Model of Disease Causation.

The study

The case study to which we refer was performed in Australia in 1979; a detailed account of this study has been published (Beck, 1985). Australia at the time of the study was inhabited by a population of 15 million (Australian Bureau of Statistics, 1989), 160 000 of whom identified themselves as Aboriginal (Australian Bureau of Statistics, 1986). The majority of the European or non-Aboriginal population resides in the southeastern section of the country. This division is so marked that Australia can be considered to consist of two sections: 'industrial' Australia with its predominantly European, non-Aboriginal population, and 'colonial' Australia, where the Aboriginal population constitutes a considerable proportion of the resident population (Fig. 9.3). This division has been described by Rowley (1970) in the 1960s in terms of 'settled' and 'colonial' Australia, and persists to this day. Conversely, two thirds of the Aboriginal

Fig. 9.3 The distinction between 'colonial' and 'industrial' Australia, and the distribution of localities within 'colonial' Australia (Heppell, 1979, p. 4).

population live in colonial Australia and the remainder in industrial Australia. The Aboriginal population in industrial Australia either live in urban centres or on rural settlements, while in colonial Australia they also live in a number of additional locations. Because of a decentralisation movement that occurred in rural government settlements and missions, in addition to a movement away from pastoral properties, many Aborigines have in the last few decades moved to live on outstations and in town-camps (Beck, 1985). Alice Springs, which is located in a semi-arid section of the Northern Territory, is surrounded by a number of Aboriginal settlements, missions and pastoral properties and has been a major recipient of this rural–urban flux. Although few families or individuals have actually moved into the town proper, the town-camps have increased significantly in size over the past few decades (Beck, 1985, pp. 13–21). The town-camps at the time of the study were distributed primarily on the perimeter of the town. A population census, held in the town-camps prior to the study, identified 1016 residents distributed over 35 camps, 26 of which had children as permanent residents (Australian Department of Aboriginal Affairs, 1979) (Fig. 9.4).

The aim of the study was to investigate the interaction between biological and cultural factors influencing the morbidity parameters of the Aboriginal children living in the Alice Springs town-camps. A diverse range of research methods was used: an anthropometric and morbidity survey was performed on all children living in the town-camps (Beck, 1985). Using these parameters as dependent variables, three camps were identified in which data on social and economic aspects of their inhabitants' lives were collected. Genealogies were drawn up, a semi-structured questionnaire was devised for the sociocultural survey (Beck, 1985) and observational methods were also used. Data collected included demographic characteristics of the camps' inhabitants, descriptions of family dynamics, ceremonial involvement and the operative medical belief system, and health services utilisation. These were complemented by data on income and expenditure, dietary patterns and material possessions.

Using Waterloo's (1976) criteria for undernutrition transposed onto NCHS standards (United States National Center for Health Statistics, 1976), those children on or below the fifth percentile for height-for-age were defined as having chronic undernutrition. Similarly those on or below the tenth percentile for weight-for-age were defined as having acute undernutrition, while being on or below the tenth percentile for weight-for-height was indicative of an inadequate energy intake. Fig. 9.5 to 9.7 depict the results of this cross-sectional survey performed on the children. While there is evidence of significant over-representation for both height-for-age and weight-for-age for the lower percentiles of the study population

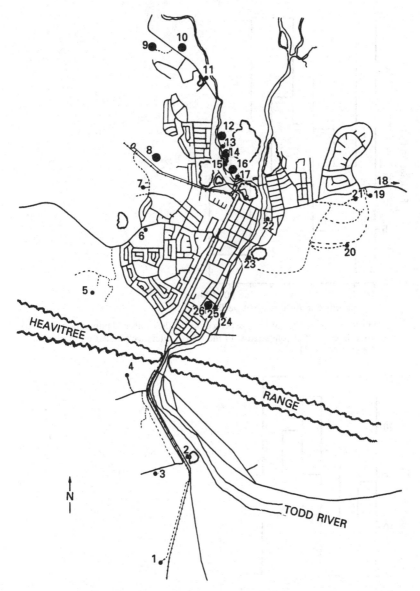

Fig. 9.4 The distribution of town-camps (circles) in and around Alice Springs included in the survey (Beck, 1985, p. 16).

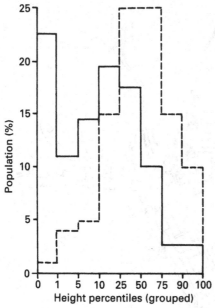

Fig. 9.5 Grouped height-for-age percentiles for the total child population surveyed (solid line; $n = 200$) compared with the NCHS reference population (dashed line) (Beck, 1985, p. 31).

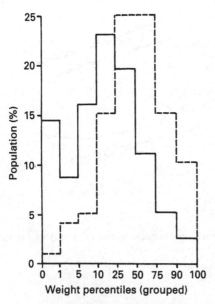

Fig. 9.6 Grouped weight-for-age percentiles for the total child population surveyed (solid line; $n = 200$) compared with the NCHS reference population (dashed line) (Beck, 1985, p. 33).

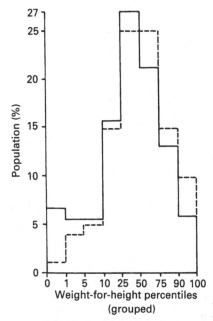

Fig. 9.7 Grouped weight-for-height percentiles for the total child population surveyed (solid line; $n = 185$) compared with the NCHS reference population (dashed line) (Beck, 1985, p. 35).

compared with the reference population, this is not observed for weight-for-height. This suggests that, although the energy intake of these children was adequate, the level of protein intake was inadequate. Of all the children surveyed, 49% had signs of undernutrition; using height-for-age and weight-for-age as nutritional indices, 10% had evidence of chronic undernutrition, 15.5% had evidence of acute undernutrition and 23.5% had evidence of both acute and chronic undernutrition. Using height-for-age and weight-for-height as nutritional indices, 30% had evidence of chronic undernutrition, 14% had evidence of acute undernutrition and 5% had evidence of both acute and chronic undernutrition (Beck, 1985).

Several indicators of disease were examined for. The prevalence of infectious diseases was found to be high in the town-camp children. Gastrointestinal and respiratory infections predominantly affected the infants and younger children: eye, skin and dental status deteriorated with age, but ear problems remained constant over time. Significant inverse relationships were observed between nutritional indices and specific infectious diseases, especially gastrointestinal disturbances, respiratory infections, middle ear disease and ocular infections (Beck, 1985). An overall index of infective load was created and named Morb Score (Beck,

138   E. J. Beck

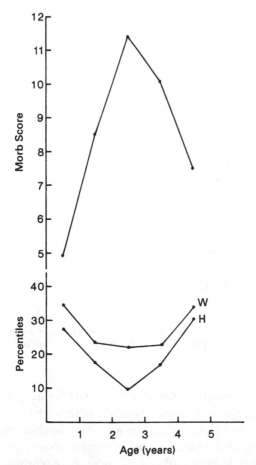

Fig. 9.8 Synergistic interaction between undernutrition, mean height-for-age (H) and weight-for-age (W) percentiles (NCHS standards), and infectious load, mean Morb Score (Beck, 1985, p. 56).

1985). Mean height-for-age, weight-for-age and Morb Scores were calculated for each age category. In the population aged under five, evidence of the synergistic action between undernutrition and infection was observed (Fig. 9.8) in terms of significant inverse relationships between mean Morb Score and mean height-for-age and weight-for-age levels, respectively ($r = -0.13$, $p < 0.05$, and $r = -0.15$, $p < 0.05$) (Beck, 1985, pp. 54–7). No differences were observed in height-for-age percentiles when the children were analysed in terms of their lingusitic background, comprising Walpiri, Eastern Aranda or Western Aranda–Luritja groups. This was indicative of a similar past environmental background, as Keats (1976) had already demonstrated the genetic similarity between these linguistic groups. Difference were observed in terms of weight-for-age and weight-for-height, i.e. acute nutritional indices caused by contemporary environmental differences (Beck, 1985). Subsequent to indirect standardisation of height-for-age, weight-for-age and Morb score for the larger camps, three camps were selected in which the social and economic investigations were performed.

Camps 9 and 18, consisting of Walpiri and Eastern Aranda linguistic groups respectively, had 'better' nutritional indices compared with camp 8, 8, which consisted predominantly of Western Aranda–Luritja people (Beck, 1985). The physical conditions in which the town-camp dwellers lived at the time of the survey were similar in the three camps. All lived in humpies or tents with access to communal toilets and laundries. Subsequent to the completion of the study, the inhabitants of all three camps moved into European-type dwellings, which had been designed to local Aboriginal requirements.

Traditional Aboriginal society was characterised by three inter-dependent characteristics (Beck, 1985):

(1) a well-developed kinship system;
(2) a hunter and gatherer mode of production;
(3) ceremonial cults linking people to their natural environment and social structure.

Normative behaviour among the town-camp inhabitants continued to be strongly influenced by their kinship system. The degree of adherence to this in individual camps reflected the duration of European contact history: the linguistic groups living in camps 8 and 18 had a substantially longer European contact history than the linguistic group living in camp 9. Of the 51 identified marriages, seven marriages among camp 8 and 18 inhabitants contravened subsection guidelines, while one marriage in camp 9 was found to be polygamous. Similarly, while a reduction in age differential between spouses was observed among camps 8 and 18 marriages, this was not observed in camp 9. Avoidance behaviour seemed to be more strictly adhered to in camp 9, less so in camp 8 and least in camp 18. Knowledge of the subsection system, especially among the younger members living in the camps, also reflected this pattern; it was best known by the inhabitants of camp 9, less so by those of camp 8 and least among camp 18 inhabitants. Ceremonial life continued among most town-camp dwellers and was valued in all camps, although high-density living had forced the adoption of new practices. While disease causation was still primarily expressed in traditional terms and traditional healers continued to practise in the town-camps, an interplay between traditional and allopathic services had come to exist (Beck, 1985).

The mode of production of the town-camp dwellers consisted of the kinship system or the utilisation of other economic resources, especially the European cash economy. The importance of the kinship system in terms of operating as part of the mode of production has been observed in other societies (Godelier, 1975). Food continued to be obtained through hunting or gathering in the camps. Because of the relative denudation of the area immediately surrounding Alice Springs, individuals had to rely on the availability of private transport, especially cars, in order to continue this practice. Alternatively, these foods were obtained when camp dwellers visited their respective parent communities. Monetary income could be obtained through the kinship network, as described by Hamilton (1971), or directly through the European cash economy. The philosophy that persisted towards the acquisition of material goods and money continued to be characterized by 'limited and concrete objectives' (Collmann, 1979, p. 131). The inhabitants had short-term monetary objectives and were uninterested in creating a material surplus; they seemed more interested in maintaining and extending social relationships. All camps were heavily reliant on the European cash economy, primarily through the social security system and to a lesser extent paid employment. Camp 18 was most successful in utilizing the options provided through both social security and paid employment. Though camp 8 had the highest level of paid employ-ment, its inhabitants did not utilise the social security system to the same extent as inhabitants of camps 9 and 18. The inhabitants of camp 9 operated primarily through utilising the social security system of the European cash economy in conjunction with their close kinship networks. These networks were also maintained through the extensive social exchange that existed between the town-camps and their respective rural parent communities.

Expenditure in the camps was mainly channelled into two items, food and transport. The food consumed consisted primarily of refined carbohy-drate items, fresh or canned meat, with relatively little fresh vegetables or fruit. When possible this was supplemented by food derived from hunting or gathering. Given the importance placed on maintaining social ties through visiting relatives in and around Alice Springs, a considerable proportion of their income was spent on transport. Because of the paucity of public transport in Alice Springs, taxis were used extensively, as not many individuals possessed a car. Expenditure continued to be character-ised by a cycle of feast and famine with a fortnightly pattern, coinciding with the fortnightly issuing of social security benefits (Beck, 1985).

In summary, the nutritional status of the town-camp children was found to be poor, with 49% showing evidence of undernutrition. A high infectious disease load was observed and an inverse relationship between nutritional status and infectious disease load demonstrated. Those age groups with maximal growth retardation were found to have the greatest infectious disease load: the under-fives were most severely affected. The physical environmental factors were poor in all camps. In the selected camps, the kinship system continued to prevail. Similarly, a high level of involvement in ceremonial life was observed with adherence to a traditional framework of disease causation and management. Economic parameters were indicative of low monetary income, which had to support large extended families. Expenditure was primarily governed by food and transport requirements. Camps with 'better' nutritional indices were those willing or able to utilise more proficiently the economic avenues open to them. The social and economic conditions found to operate in the town-camps were an integral part of the social and economic conditions operative in Central Australia at large, and the poor health patterns of the children living in the town-camps were a direct consequence of the continuing process of colonisation affecting this region (Beck, 1985).

Summary

The Ecological Model of Disease Causation provides us with a useful framework whereby we can consider the interactive influence of those factors that determine disease patterns. Using height-for-age as an index of chronic undernutrition, dietary protein in our example can be considered to be the agent. Its relative deficiency constitutes a significant part of the pathophysiological mechanism that results in stunting. Protein deficiency is, however, further aggravated through the synergistic relationship between undernutrition and infection. Infectious disease agents, viral or bacterial, in this context act as cofactors. Environmental factors can enhance the transmission of these agents to susceptible hosts. This can also occur through physical factors, like poor sanitation or overcrowding, or cultural factors including specific behavioural practices. At the host level, genetic and cultural factors need to be considered. While older children can display much genetic variation in stature, both in young children and at a population level, these genetic influences are less important. Cultural factors in this context relate primarily to the dietary habits of the individual, especially during periods of rapid growth. These are related to dietary preferences but also to the ability to purchase certain food items. Food preferences and access to these items also operate at a population level and are therefore dependent on the social and economic context which relates to the particular community. When a hunter–gatherer community is

142   *E. J. Beck*

engulfed by a technologically superior and advancing population, the latter drastically affects the context of the former's existence, including their ability to sustain themselves.

The model allows us to interpret the importance of an observed phenomenon within its biomedical and social contexts. Being 'short' is not necessarily disadvantageous or harmful. If one is 'stunted', however, this is associated with significant morbidity or mortality and is harmful to that individual. Finally, the model not only provides us with a framework whereby we can observe and analyse various morbidity or mortality patterns in populations, but it also provides us with a comprehensive framework in which to design and evaluate appropriate intervention measures at the level of the agent(s), host and environmental factors (Beck, 1985).

References

Australian Bureau of Statistics (1986). *Year Book of Australia – 1986.* Canberra: Bureau of Statistics.
Australian Bureau of Statistics (1989). *Year Book of Australia – 1989.* Canberra: Bureau of Statistics.
Australian Department of Aboriginal Affairs (1979). *Aboriginal population census – Alice Springs, 1979.* Unpublished data, Alice Springs.
Babu, D. S. & Chuttani, C. S. (1979). Anthropometric indices independent of age for nutritional assessment in school-children. *British Journal of Preventive and Social Medicine* **33**, 177–9.
Barker, W. H. (1975). Perspectives on acute enteric disease: epidemiology and control. *Pan American Health Organisation Bulletin* **9**, 148–56.
Beck, E. J. (1985). *The Enigma of Aboriginal Health: interaction between biological, social and economic factors in Alice Springs town-camps.* Australian Institute of Aboriginal Studies, Canberra.
Beck, E. J., Donegan, C., Cohen, C. S., Kenny, C., Moss, V., Underhill, G. S., Terry, P., Jeffries, D. J., Pinching, A. J., Miller, D. L., Cunningham, D. G. & Harris, J. R. W. (1989). Risk factors for HIV-1 infection in a British population: lessons from a London sexually transmitted diseases clinic. *AIDS* **3**, 533–8.
Blaxter, M. (1990). *Health and Lifestyles.* London: Tavistock/Routledge.
Cheek, D. B. (1968). *Human Growth.* Philadelphia: Lea and Febiger.
Collmann, J. (1979). *Burning Mount Kelly: Aborigines and the administration of social welfare in Central Australia.* PhD thesis, University of Adelaide.
Godelier, M. (1975). *Modes of production, kinship and demographic structure.* In *Marxist Analyses and Social Anthropology* (ed. M. Bloch), pp. 3–27. London: Malaby Press.
Hamilton, A. (1971). *Socio-cultural factors in health among the Pitjantjatjara-a preliminary report.* Department of Anthropology, University of Sydney.
Heppell, M. (1979). *A Black Reality: Aboriginal camps and housing in remote Australia,* Australian Institute of Aboriginal Studies, Canberra.

Keats, B. J. B. (1976). *Genetic aspects of growth and of population structure in indigenous people of Australia and New Guinea.* PhD thesis, Australian National University, Canberra.

Kelman, S. (1975). The social nature of the definition problem in health. *International Journal of Health Services* **5**, 625–42.

Moravcsik, J. (1976). Ancient and modern conceptions of health and medicine. *Journal of Medicine and Philosophy* **1**, 337–48.

Riese, W. (1953). *The Conception of Disease: its History, its Visions and its Nature.* New York: Philosophical Library.

Rose, G. (1985). Sick individuals and sick populations. *International Journal of Epidemiology* **14**, 32–8.

Rowley, C. D. (1970). *The Remote Aboriginies.* Canberra: Australian National University Press.

Susser, M. (1989). Epidemiology today: a thought-tormented world. *International Journal of Epidemiology* **18**, 481–8.

Tanner, J. M. (1976). Population differences in body size, shape and growth rate: a 1976 view. *Archives of Disease in Childhood* **51**, 1–2.

Townsend, P. & Davidson, N. (1986). *Inequalities in Health: the Black Report* Harmondsworth: Penguin Books.

United States National Centre for Health Statistics (1976). N.C.H.S. Growth Charts, 1976. *United States Department of Health, Education and Welfare, Monthly and Vital Statistics Report* **25**, no. 3 (supplement).

Waterloo, J. C. (1976). *Classification and definition of protein-energy malnutrition.* In *Nutrition in Preventive Medicine* (ed. G. H. Beaton & J. M. Bengoa), pp. 530–55. Geneva: World Health Organisation.

10 Selection for rural-to-urban migrants in Guatemala

H. KAPLOWITZ, R. MARTORELL AND P. L. ENGLE

Previous studies of rural-to-urban migration have led to difficulties in distinguishing migration antecedents from migration consequences. Failure to control for key factors such as migration history, age at migration, and duration of migration measures make results difficult to interpret. A prospective study conducted in rural Guatemala allowed for comparisons on a large and diverse set of anthropometric, psychological and socioeconomic variables between eventual urban migrants and rural sedentes as early as age 3 years. Results indicate that eventual migrants may have been absent from the village more often than sedentes even at this young age, suggesting that these children may have been part of families with a history of circular migration. Evidence for selective migration (i.e. migration of individuals who are in some way 'better off' than the originating population) was found in two of the four study villages, where eventual migrants were larger and of better socioeconomic status at age 3 than sedentes. Few differences were found in the other two villages. Differences generally persisted subsequent to migration. Patterns seen are indicative of the economic and social systems characteristic of the individual villages: the two villages in which selective migration was seen offer little to deter individuals from leaving, while the other two villages provide greater economic opportunity for residents. Rural regional differences should be carefully considered in future studies of rural-to-urban migration.

Introduction

The migration literature has suggested that migrants are not representative of the rural areas from which they originate, i.e. that they are 'better off' or positively selected, and that it is the taller and presumably healthier individuals from the rural areas who eventually migrate (Kaplan, 1954; Illsley et al., 1963). Examples of studies specific to Latin America include a comparison of Mexican schoolchildren which found that urban, squatter settlement children were taller than rural Zapotec children (Malina et al., 1981), and a study of children from another urban squatter settlement on

144

the periphery of Guatemala City who were found to be slightly larger than children from rural areas (Johnston *et al.*, 1985). The urban groups in both of these studies were composed largely of migrants from rural areas. Bogin & MacVean (1981) compared lower socioeconomic status (SES) urban schoolchildren who had city-born parents with those with rural-born parents, and found that children of migrant parents were smaller than those of city-born parents. However, these studies and others are difficult to interpret without data on age at migration and duration of time spent in the new environment, often not available and usually combined with a lack of pre-migration measurements. It is not clear whether differences between migrants and sedentes existed before migration, and perhaps affected the likelihood of the subsequent migration, or whether they developed subsequent to migration, perhaps resulting from exposure to an improved environment.

Much of the economic and social literature supports selectivity of migrants based on SES (Rogers & Williamson, 1982; Macbeth, 1984; Bogin, 1988). In developing countries where growth status is so closely tied to poverty, there may be little difference between selectivity for growth and for SES, as greater SES will usually result in better child growth.

Selectivity may also be related to psychological factors. Researchers of 'migration psychology' have based their analyses on decision process models: a chain of psychological and social factors interact to result in the decision to migrate or not (Wolpert, 1965; Richardson, 1974; Fawcett, 1986). These factors may involve pressure from family members to go or stay, education of both the potential migrant and his or her parents, marital status, and/or differences in value-expectancies (Fuller *et al.*, 1986; De Jong *et al.*, 1986; Sly & Wrigley, 1986).

The most direct approach to unravelling the issue of selectivity is through longitudinal studies including before-and-after migration measurements (Wessen, 1971). Only a few such studies exist, particularly for Latin America. None found evidence of selective migration for physical characteristics (Lasker, 1952, 1954; Lasker & Evans, 1961; Malina *et al.*, 1982).

However, longitudinal studies of migrant selectivity based on psychological or social factors have found evidence of selection, although none have been conducted in Latin America. For example, the decision to migrate has been positively linked with measures of value-expectancy and education (Fuller *et al.*, 1986; Sly & Wrigley, 1986), family pressure to move, being single, and having a network of relatives outside the originating village (De Jong *et al.*, 1986), and negatively linked to changes in job opportunities or family relationships (Gardner *et al.*, 1986; Sly & Wrigley, 1986). The decision to migrate has also been related to the social aspirations of the migrant and/or his or her family (Teller, 1973; Goldstein, 1978). To our knowledge, no studies exist of cognitive test performance in migrants prior to migration.

Recent research in Guatemala provides a unique opportunity to address the issue of selection of rural-to-urban migrants. Measurements on a large and diverse set of growth and development variables, in addition to an unusually large battery of socioeconomic measures, were taken during a longitudinal study of early child growth in rural Guatemala. To our knowledge, no other study has such extensive physical, psychological and socioeconomic data on eventual migrants and their rural village peers as early as the first three years of life, presumably before the decision to migrate was made, and then again after migration.

With this in mind, these analyses were undertaken to explore the potential biological, psychological and socioeconomic differences during early childhood between eventual urban migrants and rural sedentes, and to determine whether these differences persist subsequent to the event of migration.

Methods

Data presented here come from two sources. The early childhood data originate from a longitudinal intervention study conducted by the Institute of Nutrition of Central America and Panama (INCAP) from 1969 to 1977. Four rural Ladino villages in northeastern Guatemala were identified for participation in a nutrition supplementation programme with the objective of determining the effects of improved nutrition on growth and development of young children. Two villages, Conacaste and San Juan, received a high-energy, high-protein treatment (atole), and two villages, Santo Domingo and Espiritu Santo, received a low-energy, no-protein supplement (fresco). Santo Domingo and Conacaste were larger villages of about 1000 people, whereas Espiritu Santo and San Juan were smaller villages of about 800 residents each in 1974. The study sample included children from birth to age 7 years. Additional details of the design and intervention have been provided elsewhere (Martorell *et al.*, 1980, 1982).

Anthropometric examinations were conducted every 3 months until 2 years of age, every 6 months from 2 to 4 years, and at ages 5, 6, and 7. A battery of pre-school psychological tests was administered annually to children ages 3 through 7 years. A variety of socioeconomic variables were measured in a 1974 census and through other surveys taken during the longitudinal study.

Adolescent and young adult status data come from a 1988–9 cross-sectional follow-up of these same participants, now aged 11–26 years.

Table 10.1. *Sample sizes by migration status, gender and village*

Village	Urban migrants			Rural sedentes		
	males	females	all	males	females	all
Santo Domingo	45	28	73	142	136	278
Conacaste	31	32	63	145	146	291
Espiritu Santo	22	22	44	106	106	212
San Juan	23	27	50	115	93	208
All	121	109	230	508	481	989

These individuals were located and identified for follow-up through a 1987 census. Adolescent and young adult data included a range of anthropometric and psychological measures, retrospective life histories, socioeconomic measures, and income and wealth information. A selection of these measures are included here.

Attempts were made to locate those individuals in the above sample who had migrated to Guatemala City, and additional data were collected on this subsample through a migration survey that characterised aspects of their migration. A rural-to-urban migrant was defined here as anyone from the four villages who had participated in the intervention study at some time between birth and age 7 years, and who either (1) was a current urban migrant who completed a migration survey ($n = 165$), or (2) was identified through his or her retrospective life history as having spent one or more years in an urban area, including migrations to other countries, at any age subsequent to 7 years ($n = 65$). Of these additional 65 individuals, about 95% had previously migrated to Guatemala City. These two subsamples were similar on all early childhood variables included in these analyses. This migrant sample represents about 70% of those who migrated to urban areas from the four villages. The remaining 30% is composed of migrants currently beyond the study limits (e.g. to remote areas of Guatemala or to other countries) or Guatemala City migrants who could not be located. Non-migrants (sedentes) were identified through the retrospective life histories as those individuals who spent no more than one month consecutively outside their village. There were 989 sedentes from the four villages (Table 10.1). Individuals excluded from these analyses included rural-to-rural migrants, individuals who spent more than one month but less than one year in the city, and those individuals who were lost to follow-up.

Some children may have left the village for short, temporary intervals;

for example, to accompany their parents on seasonal employment migrations. Since prior mobility has been related to the migration decision (Fuller *et al.*, 1986), it is important to know if eventual urban migrants were more mobile than sedentes, even at this early age. Measures of residency (days present in the village) were calculated from child morbidity surveys administered bi-weekly to the mothers. The variables DAYS1ST, DAYS2ND and DAYS3RD were calculated by year of life (birth to 12, 13–24 and 25–36 months) in order to avoid bias due to staggered study entry and departure. Since refusal to participate in these morbidity surveys was quite low (no more than a handful of families per village), missing values indicate that the child, or at least the child's mother, was absent from the village during the given period. Eventual migrants were then compared with non-migrants on a number of early childhood measures, including dietary, biological, psychological and socioeconomic variables.

Biological measures included length (LN36) and head circumference (HDCIR36), measured in centimetres within 7 days of reaching 36 months of age, as well as maternal height (MHT; cm). Psychological measures included two factors from a factor analysis of the pre-school battery of measurements at age 7 years, as cognitive measures from earlier ages are generally less predictive of subsequent behaviour and may have lower test–retest reliability. In this case, data at 7 years also provided a larger sample and were the closest to school age. Details of the individual tests and the factor analysis are provided by Pollitt & Gorman (1990). In brief, the first factor was a general perceptual–organisational and verbal factor (VERBAL84). Tests with the largest loadings are both the perceptual organisation and verbal ability tests. The second factor (SHMEM84) loaded most heavily on digit and sentence memory. A 'naming' test (number of items named correctly; NAME84) was also included separately in addition to its inclusion in the factor analysis, as it was the only measure of verbal output. Because this test was first administered in 1969, whereas others in the battery were initiated in 1971, more subjects were available for analysis using this test. These early test performance measures are related to schooling performance in this population (Irwin *et al*, 1978).

Socioeconomic measures included data from the 1974 census and parental tests and questionnaires. An index of housing quality (HOUSE74) was derived from a factor analysis of physical characteristics of the home, including materials, presence and type of sanitation facilities, and separate cooking areas. Mothers were asked whether they wore a sweater or shoes (CLOTH74), another indication of wealth, and also how many individuals were responsible for teaching the child numbers and colours, for example (TEACH74). These measures were all standardised within village to control for differences between the villages and allow for comparisons within villages. Household stimulation (HSTIM) was measured by summing standardised scores for the amount of intellectually stimulating material that could be found in the home. Also included were measures of maternal vocabulary (VOCABM) and literacy (LITERATEM), the number of school grades passed by the parents (GRADESM, GRADESF), and a measure of maternal modernity (MODERNITY). These measures have been described in detail elsewhere (Engle *et al.*, 1979, 1983; Irwin *et al.*, 1978). In addition, measures of maternal involvement in the child's social system (e.g. did the family make or buy and maintain toys for the child or display the child's work on the walls?; INVOLVEM) and the father's social involvement in the child's life (amount of time spent with the pre-school child during the week and on weekends, measured on a 0–5 scale; INVOLVEF) were also included.

Comparisons were made on a number of post-migration variables in order to determine whether antecedent differences persisted subsequent to migration. Follow-up variables included height (HT; cm) and head circumference (HDCIR; cm), body mass index (BMI; $kg\,m^{-2}$), fat-free mass (FFM; kg), literacy (LITERATE; 1 = illiterate to 4 = able to read and write), school grades passed by the participant (GRADES), and an index of factor scores describing 1987 physical characteristics of the home (HOUSE87).

Analysis of variance was used to assess differences between eventual migrants and sedentes in early childhood. Separate models were run for each response variable described above. Independent variables included migration status (0 = sedente, 1 = eventual urban migrant), gender (0 = males, 1 = females), and the interaction between the two. The interaction term was included as men and women have different patterns of migration in Latin America.

Balán (1983) and others (Findley, 1977) have suggested that migration is a result of changes in the economic opportunities in the rural areas. The four villages are known to differ substantially on socioeconomic variables that have been related to the decision to migrate (Pollitt *et al.*, 1990); all analyses were therefore village-specific.

Analysis of covariance was used to compare urban migrants to rural sedentes at follow-up (post-migration). Again, migration status, gender, and the interaction term were included as the independent variables (as above). In order to age-adjust outcomes of interest during adolescence and adulthood, it was necessary to restrict the sedente sample to the age range of the migrants, who were on average 3–4 years older than the sedentes. This entailed excluding cases from the lower end of the sedente age range. Age at follow-up was included as a covariate in all post-migration models.

A *p* value of 0.10 was chosen as the criterion for significance of the interaction term, based on a recommendation by Cohen (1977), who suggests that conventional criteria are too restrictive for judging interactions.

Results

Early childhood comparisons

Results of the ANOVA models are presented in Tables 10.2 – 10.5; models with significant migrant status by gender interactions are shown stratified by gender in Table 10.6. It is clear that eventual migrants had a consistent, and often statistically significant, tendency to be absent from the village during early childhood more often than the sedentes (DAYS1ST, DAYS2ND, DAYS3RD).

There was a strong but non-significant tendency ($p = 0.056$) for eventual migrants from Santo Domingo to be longer at age 3 than their non-migrant counterparts. Both male and female eventual migrants from Santo Domingo also showed a strong and highly significant tendency to be larger in head circumference and maternal height. A significant migrant status by gender interaction for length at age 3 suggests a similar trend in males from San Juan (Table 10.6). There was a tendency for head circumference to be greater for eventual migrant males from Conacaste ($p = 0.058$), but there were no other anthropometric differences in Conacaste or Espiritu Santo. Significant differences were about two centimetres in all variables, a biologically important difference.

Results for the psychology measures indicate that eventual migrants from Santo Domingo scored an average of two more items correct on the naming test, on the order of half a standard deviation, and that factor scores for the verbal tests were consistent with this pattern, but not significant. A significant gender by migration status interaction in the short-term memory factor indicates that female migrants from Espiritu Santo scored significantly higher than sedentes (Table 10.6). Other differences in Espiritu Santo and San Juan were non-significant and inconsistent. Differences in Conacaste consistently favoured eventual migrants, but were never statistically significant.

Eventual migrants from Santo Domingo also scored significantly better on 8 of the 11 socioeconomic variables. Exceptions were grades passed by the mother and father and mother's modernity (although all variables but father's schooling were consistent with this trend). Migrants from San Juan scored significantly higher on 6 of the 11 SES measures, including HOUSE74, CLOTH74, TEACH74 (for males only), HSTIM, INVOLVEF, and MODERNITY. INVOLVEM approached significance

Table 10.2. *Comparison of eventual migrants and rural sedentes during early childhood: Santo Domingo*

Variable	Eventual migrants			Rural sedentes			p values[a]		
	n	mean	s.d.	n	mean	s.d.	migrant	gender	migrant by gender
Residency									
DAYS1ST	20	232.70	124.77	149	316.01	65.64	0.002	0.000	0.001
DAYS2ND	25	269.08	120.56	135	334.12	59.71	0.000	0.402	0.478
DAYS3RD	26	317.23	76.11	123	335.28	61:86	0.198	0.842	0.872
Anthropometry									
LN36	30	85.71	3.28	138	83.94	4.47	0.056	0.005	0.837
HDCIR36	29	47.64	1.38	138	46.82	1.40	0.002	0.000	0.880
MHT	42	151.55	4.50	247	149.72	4.72	0.014	0.479	0.438
Psychology									
VERBAL84	17	0.17	1.01	49	−0.17	1.13	0.220	0.939	0.454
SHMEM84	17	−0.06	0.94	49	0.01	0.92	0.527	0.251	0.382
NAME84	29	25.55	3.30	89	23.11	4.44	0.007	0.512	0.601
Socioeconomic									
HOUSE74	46	3.45	0.77	156	2.96	0.73	0.000	0.381	0.879
CLOTH74	60	22.73	76.65	246	−10.90	80.12	0.004	0.999	0.955
TEACH74	60	38.10	67.14	246	0.20	81.28	0.002	0.134	0.331
HSTIM	34	177.65	305.00	213	33.41	268.03	0.007	0.613	0.553
INVOLVEM	33	83.91	151.58	135	−21.81	185.87	0.003	0.888	0.612
INVOLVEF	34	4.65	0.73	134	4.30	0.99	0.052	0.722	0.626
VOCABM	33	9.30	3.05	135	6.16	3.51	0.000	0.953	0.471
LITERATEM	34	2.68	1.32	213	2.10	1.24	0.017	0.717	0.690
GRADESM	30	1.73	1.62	212	1.33	1.40	0.179	0.746	0.578
GRADESF	31	1.48	1.48	192	1.52	1.89	0.978	0.354	0.543
MODERNITY	33	66.70	23.58	135	59.31	25.19	0.166	0.599	0.377

[a] p Values for given ANOVA terms. See text for further explanation.

($p = 0.067$). Significant differences in Conacaste were limited to HOUSE74 and INVOLVEF, for which migrants had higher scores. No SES differences were found for Espiritu Santo, although HSTIM and TEACH74 approached significance. SES differences between migrants and sedentes in Conacaste and Espiritu Santo were also inconsistent, i.e. there was no evidence of any trend.

Table 10.3. *Comparison of eventual migrants and rural sedentes during early childhood: Conacaste*

	Eventual migrants			Rural sedentes			p values[a]		
Variable	n	mean	s.d.	n	mean	s.d.	migrant	gender	migrant by gender
Residency									
DAYS1ST	12	258.08	109.65	135	308.41	70.6	0.169	0.016	0.028
DAYS2ND	12	261.25	122.88	135	317.05	89.53	0.045	0.710	0.578
DAYS3RD	12	299.67	113.46	114	327.26	72.70	0.104	0.027	0.007
Anthropometry									
LN36	20	86.76	4.41	120	87.41	3.67	0.320	0.078	0.664
HDCIR36	20	48.12	1.43	120	47.78	1.13	0.782	0.000	0.029
MHT	46	150.58	5.81	240	150.41	5.35	0.849	0.290	0.135
Psychology									
VERBAL84	19	0.40	0.71	53	0.21	1.08	0.449	0.283	0.497
SHMEM84	19	0.08	1.14	53	−0.38	0.75	0.138	0.708	0.442
NAME84	38	25.26	3.37	81	24.77	3.06	0.436	0.690	0.184
Socioeconomic									
HOUSE74	43	2.97	0.85	158	2.61	0.89	0.024	0.782	0.479
CLOTH74	54	17.04	62.50	229	−0.32	80.24	0.131	0.416	0.416
TEACH74	54	20.06	92.02	229	11.36	71.92	0.458	0.272	0.823
HSTIM	35	62.89	251.66	178	68.40	233.12	0.873	0.926	0.434
INVOLVEM	31	−40.03	204.71	118	−10.59	181.25	0.403	0.745	0.536
INVOLVEF	35	3.86	1.54	132	3.17	1.53	0.025	0.463	0.131
VOCABM	32	5.28	2.87	119	5.39	3.10	0.710	0.201	0.107
LITERATEM	36	1.67	1.04	181	1.76	1.17	0.808	0.070	0.034
GRADESM	34	0.94	1.01	175	1.25	1.36	0.222	0.959	0.327
GRADESF	29	1.38	1.93	153	1.06	1.38	0.557	0.131	0.191
MODERNITY	31	54.90	28.62	119	48.18	24.50	0.244	0.815	0.235

[a] p Values for given ANOVA terms. See text for further explanation.

Adolescent and young adult (post-migration) comparisons

Age-adjusted post-migration means for urban migrants and sedente males and females are shown in Tables 10.7 and 10.8, respectively. The strong differences seen during early childhood in Santo Domingo appear to persist into adolescence and young adulthood, with the exception of female height for which differences are attenuated. There is a slight but consistent trend for migrants from the other villages to be slightly better off on all variables, with the exception of male stature and HOUSE87 in Espiritu Santo. Migrants, particularly female migrants, tend to be better educated in all villages.

Table 10.4. *Comparison of eventual migrants and rural sedentes during early childhood: Espiritu Santo*

| | Eventual migrants | | | Rural sedentes | | | p values[a] | | |
	n	mean	s.d.	n	mean	s.d.	migrant	gender	migrant by gender
Residency									
DAYS1ST	11	226.09	125.54	121	299.82	77.22	0.005	0.938	0.613
DAYS2ND	12	256.58	143.68	119	333.85	60.63	0.000	0.945	0.902
DAYS3RD	14	307.71	87.80	98	339.95	63.55	0.077	0.677	0.388
Anthropometry									
LN36	19	84.89	4.04	114	85.04	3.31	0.898	0.008	0.419
HDCIR36	19	46.09	0.98	113	46.21	1.40	0.828	0.000	0.091
MHT	33	146.28	4.61	200	147.48	4.24	0.144	0.382	0.911
Psychology									
VERBAL84	13	−0.14	0.95	37	−0.12	0.78	0.797	0.483	0.135
SHMEM84	13	0.75	0.82	37	−0.14	1.16	0.023	0.868	0.042
NAME84	27	22.19	3.22	61	22.31	3.66	0.688	0.749	0.043
Socioeconomic									
HOUSE74	32	2.39	0.79	119	2.34	0.85	0.735	0.471	0.281
CLOTH74	38	−24.95	70.86	194	−4.03	70.25	0.115	0.059	0.029
TEACH74	38	21.13	69.19	194	−3.45	78.93	0.069	0.498	0.367
HSTIM	26	−49.00	231.14	179	−148.34	250.36	0.058	0.631	0.746
INVOLVEM	26	64.69	260.13	109	12.01	165.88	0.201	0.797	0.474
INVOLVEF	25	2.96	1.43	113	2.57	1.22	0.136	0.976	0.167
VOCABM	26	5.85	3.92	109	5.94	3.29	0.896	0.793	0.578
LITERATEM	26	1.96	1.31	180	1.97	1.26	0.970	0.388	0.246
GRADESM	24	1.13	0.95	179	1.44	1.46	0.326	0.597	0.850
GRADESF	18	1.67	1.88	161	2.44	1.90	0.088	0.414	0.991
MODERNITY	26	42.88	26.49	109	49.05	27.41	0.303	0.971	0.487

[a] p Values for given ANOVA terms. See text for further explanation.

Discussion

The main objective of these analyses was to distinguish between anteced-ents and/or determinants of rural-to-urban migration and its results or consequences.

Examination of the average number of days individuals were present in their villages between birth and age 3 years indicates that eventual migrants, or at least their mothers, were absent from the village significant-ly more often than sedentes. Therefore, early childhood comparisons may be indicative of differential mobility patterns. That is, perhaps the fact that eventual migrants were better off in two villages at age three was a

Table 10.5. *Comparison of eventual migrants and rural sedentes during early childhood: San Juan*

Variable	n	mean	s.d.	n	mean	s.d.	migrant	gender	migrant by gender
		Eventual migrants			Rural sedentes			p values[a]	
Residency									
DAYS1ST	14	251.86	101.84	120	300.76	69.63	0.306	0.003	0.048
DAYS2ND	17	272.59	129.15	108	319.76	74.48	0.107	0.327	0.573
DAYS3RD	22	284.27	111.27	93	313.97	86.47	0.243	0.552	0.786
Anthropometry									
LN36	22	86.80	3.41	104	86.36	3.42	0.441	0.025	0.078
HDCIR36	22	47.39	0.98	104	47.69	1.36	0.504	0.002	0.496
MHT	37	148.23	4.44	191	148.42	5.23	0.792	0.868	0.407
Psychology									
VERBAL84	13	−0.17	0.95	29	0.13	1.02	0.317	0.264	0.947
SHMEM84	13	0.03	0.80	29	−0.36	1.14	0.344	0.178	0.355
NAME84	23	22.96	4.89	47	24.07	4.28	0.283	0.559	0.752
Socioeconomic									
HOUSE74	44	3.20	0.61	99	2.91	0.67	0.017	0.978	0.631
CLOTH74	39	25.15	59.90	180	−10.26	86.58	0.015	0.918	0.786
TEACH74	39	40.05	89.24	180	−4.53	81.04	0.004	0.570	0.024
HSTIM	32	39.34	298.77	150	−108.91	195.38	0.000	0.077	0.146
INVOLVEM	33	74.12	230.70	83	−7.27	213.16	0.067	0.277	0.488
INVOLVEF	29	4.52	0.95	77	4.17	1.26	0.230	0.972	0.370
VOCABM	33	6.55	3.54	83	6.10	3.21	0.507	0.515	0.370
LITERATEM	31	2.03	1.28	148	2.05	1.29	0.975	0.860	0.507
GRADESM	30	0.87	1.01	149	1.08	1.23	0.416	0.950	0.739
GRADESF	32	1.09	1.33	138	0.86	1.23	0.352	0.688	0.773
MODERNITY	31	66.45	19.23	83	53.46	21.93	0.005	0.284	0.239

[a] p Values for given ANOVA terms. See text for further explanation.

consequence of prior mobility behaviour, and yet it could still be a determinant of future migrations. It is possible that children who eventually migrate to urban centres are members of families with a history of circular migration; that is, families who repeatedly leave and later return to the village. These families may have an inherently different lifestyle, and perhaps also different attitudes and aspirations, from sedente families.

Given the large number of statistical tests performed, some significant differences would be expected to be simply due to chance. However, both male and female migrants from Santo Domingo showed a strong and consistent tendency to have been better off during early childhood than

Table 10.6. *Gender-specific ANOVAs for dependent variables with gender × migration status interactions*

Variable		Migrants			Sedentes			
		n	$\bar{x}$	s.d.	n	$\bar{x}$	s.d.	p^a
Santo Domingo								
DAYS1ST	MALES	14	195.14	131.08	76	312.64	67.38	0.000
	FEMALES	6	320.33	36.37	73	319.51	64.05	0.975
Conacaste								
DAYS1ST	MALES	8	222.00	119.49	65	305.95	74.80	0.007
	FEMALES	4	330.25	18.84	70	310.69	67.05	0.565
DAYS3RD	MALES	7	347.43	28.61	66	322.67	78.91	0.415
	FEMALES	5	232.80	156.83	48	333.58	63.46	0.006
HDCIR36	MALES	13	48.77	1.04	62	48.09	1.18	0.058
	FEMALES	7	46.93	1.31	58	47.49	0.98	0.196
VOCABM	MALES	18	6.06	3.30	61	5.30	2.89	0.346
	FEMALES	14	4.29	1.86	58	5.50	3.33	0.194
MLIT	MALES	20	1.30	0.73	88	1.80	1.19	0.077
	FEMALES	16	2.13	1.20	93	1.73	1.15	0.213
Espiritu Santo								
HDCIR36	MALES	11	46.31	0.76	64	46.87	1.28	0.168
	FEMALES	8	45.79	1.20	49	45.36	1.04	0.292
SHMEM84	MALES	6	0.34	0.80	15	0.25	1.39	0.888
	FEMALES	7	1.11	0.69	22	−0.41	0.92	0.000
NAME84	MALES	13	21.46	3.76	25	23.44	3.82	0.167
	FEMALES	14	22.86	2.60	36	21.53	3.38	0.668
CLOTH74	MALES	20	−48.95	58.37	95	−2.16	69.16	0.006
	FEMALES	18	1.72	75.42	99	−5.83	71.59	0.684
San Juan								
DAYS1ST	MALES	4	331.00	36.23	73	309.53	58.34	0.470
	FEMALES	10	220.20	103.17	47	287.13	83.09	0.031
LN36	MALES	10	88.55	3.49	60	86.53	3.31	0.080
	FEMALES	12	85.34	2.66	44	86.14	3.59	0.478
TEACH74	MALES	21	59.00	74.61	98	−15.73	80.77	0.000
	FEMALES	18	17.94	101.45	82	8.87	79.80	0.679

[a] p Value for MIGRANT status ANOVA term. See text for further explanation.

their rural counterparts on the majority of variables examined. Migrants from San Juan were also generally better off on socioeconomic measures. Few differences or trends were notable in the other two villages. In no instance were rural sedentes significantly better off than the eventual migrants. Therefore, it is reasonable to conclude that selective migration does occur under some circumstances, and that these circumstances are village-specific.

Table 10.7. *Age-adjusted least-square means for male migrants versus sedentes at follow-up (post-migration)*

Variable[b]	Eventual migrants			Rural sedentes			
	n	Ls*x*	s.e.	*n*	Ls*x*	s.e.	*p*[a]
Santo Domingo							
HT	45	160.74	1.19	114	156.95	0.74	0.008
HDCIR	45	53.16	0.24	114	52.36	0.15	0.007
BMI	45	20.73	0.29	114	19.77	0.18	0.006
FFM	45	45.93	0.88	113	42.50	0.55	0.001
LITERATE	45	1.62	0.10	110	1.65	0.07	0.811
GRADES	45	5.91	0.38	110	4.18	0.24	0.000
HOUSE87	34	4.75	0.13	79	3.72	0.08	0.000
Conacaste							
HT	30	159.74	1.43	117	160.07	0.71	0.836
HDCIR	31	53.70	0.24	117	53.09	0.12	0.026
BMI	30	20.34	0.36	117	20.08	0.18	0.514
FFM	30	45.00	1.09	117	44.64	0.54	0.768
LITERATE	28	1.42	0.17	114	1.29	0.08	0.484
GRADES	28	3.81	0.55	114	2.70	0.27	0.075
HOUSE87	24	4.19	0.24	82	3.70	0.13	0.082
Espiritu Santo							
HT	22	160.88	0.95	43	162.94	0.67	0.086
HDCIR	22	52.80	0.30	43	52.20	0.21	0.110
BMI	22	21.90	0.46	43	20.81	0.33	0.063
FFM	22	48.10	0.98	43	47.92	0.70	0.881
LITERATE	20	1.79	0.11	40	1.90	0.08	0.403
GRADES	20	6.96	0.55	40	6.27	0.38	0.311
HOUSE87	17	3.76	0.27	41	3.84	0.17	0.806
San Juan							
HT	23	154.76	2.06	91	155.52	0.95	0.748
HDCIR	23	52.76	0.32	91	52.67	0.15	0.815
BMI	23	20.07	0.41	90	19.37	0.19	0.139
FFM	23	42.12	1.38	90	41.33	0.64	0.617
LITERATE	22	1.58	0.17	88	1.58	0.08	0.244
GRADES	22	4.51	0.71	88	3.99	0.32	0.525
HOUSE87	19	3.51	0.23	53	3.44	0.13	0.793

[a] *p* Value for MIGRATION status ANOVA term.
[b] Variable names: HT, height (cm); HDCIR, head circumference (cm); BMI, body mass index; FFM; fat-free mass (kg); LITERATE, ability to read and write on a scale of 1–4; GRADES, school grades passed; HOUSE87, house physical characteristics (factor scores).

Table 10.8. *Age-adjusted least-square means for female migrants versus sedentes at follow-up (post-migration)*

Variable[b]	Eventual migrants			Rural sedentes			
	n	Lsx	s.e.	n	Lsx	s.e.	p[a]
Santo Domingo							
HT	28	149.99	1.15	95	148.86	0.62	0.389
HDCIR	28	51.42	0.28	95	50.78	0.15	0.0444
BMI	28	22.06	0.63	95	21.09	0.34	0.181
FFM	28	37.25	0.91	95	35.14	0.49	0.044
LITERATE	26	1.78	0.13	86	1.63	0.07	0.326
GRADES	26	6.07	0.47	86	3.92	0.26	0.000
HOUSE87	20	4.83	0.20	71	4.02	0.11	0.001
Conacaste							
HT	31	150.53	1.08	118	150.20	0.53	0.783
HDCIR	31	51.69	0.24	119	51.71	0.12	0.967
BMI	31	22.58	0.48	118	21.21	0.23	0.013
FFM	31	38.47	0.80	118	36.15	0.39	0.012
LITERATE	29	1.58	0.17	113	1.29	0.08	0.141
GRADES	29	3.44	0.46	113	2.58	2.77	0.099
HOUSE87	21	3.88	0.26	76	3.75	0.13	0.650
Espiritu Santo							
HT	22	149.48	0.82	65	148.22	0.47	0.187
HDCIR	22	50.61	0.27	65	50.36	0.15	0.417
BMI	22	21.98	0.71	65	22.15	0.41	0.834
FFM	22	37.00	0.88	65	35.86	0.51	0.267
LITERATE	20	2.06	0.15	61	1.59	0.08	0.007
GRADES	20	7.69	0.58	61	4.18	0.33	0.000
HOUSE87	18	3.38	0.27	61	3.43	0.14	0.863
San Juan							
HT	27	151.04	1.35	47	150.35	1.03	0.686
HDCIR	27	51.85	0.28	47	51.70	0.21	0.651
BMI	27	22.68	0.63	47	21.74	0.48	0.239
FFM	27	38.30	0.93	47	36.63	0.71	0.160
LITERATE	24	1.63	0.16	44	1.50	0.12	0.523
GRADES	24	4.88	0.57	44	3.52	0.42	0.060
HOUSE87	22	3.85	0.20	45	3.55	0.14	0.217

[a] p Value for MIGRATION status ANOVA term.
[b] For explanation of variable names, see Table 10.7.

Balán (1983) has suggested that differences among rural areas in economic opportunities and/or social structure may be related to differential migration. He cautions, however, that 'migration is often a response to societal changes over which individuals have no control', and that 'analyses that attribute migration solely to individual motivations and behaviours cannot fully grasp its causes and implications' (p. 181). Eventual migration appears to be related to SES in both Santo Domingo and San Juan. In addition, results presented for Santo Domingo suggest that selection in this village may be partly biological. In developing countries, however, growth and socioeconomic status reflect in large part the same phenomenon, i.e., poverty, making the distinction between the two 'types' of selection trivial.

Few differences, other than those presented in Table 10.7 and 10.8, were found among the four villages in terms of the current migrants' circumstances in the city. Migrants from all four villages migrated to similar areas of the city, all known to be poor with little or no access to city services, fresh water or sanitation. Seventy per cent or more moved to the city for the first time, and about 50% of females and 75% of males migrated alone. The primary reason for migration was for better jobs or salaries, followed by migration with or to be with parents. By definition, this study was limited to an age range (11–26 years) that still allowed physical growth. The average age of current migrants was about 20 years, 3 or 4 years older than the sedente sample. Duration of migration averaged about 7 years at the time of follow-up for those from the large villages, and 4 years for those from the small villages, ample time for growth to be affected in this group. Average age at migration was slightly younger for Santo Domingo, the village showing the strongest migrant–sedente differences, about 12 years versus 16–17 years for the other villages.

The only suggestion of differences among migrants from the four villages seems to be in their 'sense of permanency'. Smaller percentages of migrants from Santo Domingo and San Juan sent remittances home during their first year in the city (about 40% versus 55%), perhaps suggesting weaker ties to their villages or different motivations for migration. Over 80% of migrants from these two villages planned to stay in the city permanently versus about 50% in the other two, and about 65% of those from the village showing the greatest differences (Santo Domingo) had arranged for employment prior to arrival in the city, compared with about 45% from the other three villages. Therefore, it may be that the reasons for migration of people from the two villages with evidence of selective migration were actually quite different from those of the other two villages: they seemed to exhibit more permanent plans to make a life for themselves in the new urban environment.

The differences between migrants and sedentes are consistent with what is generally known about the four villages. There is little to keep potentially productive individuals in Santo Domingo and San Juan. Santo Domingo offers little to no wage work on a full-time, year-round basis (Pivaral, 1972; Clark, 1979). It is also closest to Guatemala City in terms of distance and accessibility. San Juan is probably the poorest of the four villages. The land is of poor quality and difficult to farm. Only 25% of male household heads were literate in 1974, compared with 32–44% in the other three villages. However, over the past few years, trucks have begun to pick up day workers for nearby farms, perhaps deterring some individuals who would have otherwise migrated, which may partly explain why the differences found were weaker than those found in Santo Domingo. Although located close to Guatamala City, it is not very accessible owing to its distance from the main road, which is accessed over poor-quality roads, making it the most isolated of the four communities.

On the other hand, Conacaste and Espiritu Santo both offer much to encourage individuals to remain in the village. The Institute for Cultural Affairs began a large development project in Conacaste shortly after the INCAP project left (West, 1986). Bank loans were made available to farmers who wanted to invest in drip irrigation; running water was installed; a pre-school was begun, which helped to free women to work; and Conacaste now produces a cash crop of tomatoes. In addition, there is a cement factory nearby that employs many of the local men. Espiritu Santo has nearby farms and tobacco plantations offering year-round wage work to both men and women, as well as a strong cottage industry, in which they make and sell baskets, fans, and other products made from palm branches. Espiritu Santo was the only village to have electricity during the study period, and it also had the highest literacy rate among male household heads (42%). It is also the village furthest from Guatemala City, as well as from the main road leading to the city, which may act as an additional deterrent to migration.

In conclusion, it is clear that characteristics of place of origin should be considered in investigations of rural-to-urban migration. It may be, for instance, that wage differentials between rural villages and the city vary, encouraging the 'best' of the rural areas with large differentials to outmigrate, leaving the villages depleted of their potentially most productive individuals (Terrell, 1984). More broadly, the complex relationship between individual characteristics or abilities that may enable individuals to migrate and the social and economic structures of their environment that may encourage or deter them from such a migration should be considered in migration models.

Circumstances within the urban environment may also differ among migrants, potentially affecting the individual's ability to benefit from the urban setting. Gender differences, although not a strong factor in these analyses, must also be considered, as males and females may be migrating for different reasons and to very different urban situations. In addition, it becomes difficult to think of migration as a single event in time, particularly in developing countries where high rates of circular or seasonal migration are common. Migration may be a lifestyle pattern for some.

Policies which have attempted to influence population distribution have tended to be limited in scope and have had little success, perhaps because they generally focus on the urban environment rather than on the originating rural areas. Investments in improving employment opportunities and infrastructure, for example, in the rural areas, particularly targeting those areas that provide fewer opportunities, may have a greater affect on the stemming of rural-to-urban migration in the Third World.

Acknowledgements

The authors would like to express their appreciation to all members of the Guatemala Oriente Study research team, particularly to Drs Kathleen Gorman and Ernesto Pollitt for providing the results of the factor analysis on the pre-school psychology battery, as well as advice about the use of psychological variables, and to Dr Juan Rivera for his helpful suggestions regarding analytic strategy and interpretation. This research was supported by grant HD22440 from the National Institutes of Health.

References

Balán, J. (1983). Agrarian structures and internal migration in a historical perspective: Latin American case studies. In *Population Movements: Their Forms and Functions in Urbanization and Development* (ed. P. A. Morrison), pp. 151–85. Liège, Belgium: Ordina Editions.

Bogin, B. (1988). Rural-to-urban migration. In *Biological Aspects of Human Migration* (ed. C. G. Mascie-Taylor & G. W. Lasker), pp. 90–129. Cambridge University Press.

Bogin, B. & MacVean, R. B. (1981). Bio-social effects of migration on the development of families and children in Guatemala. *American Journal of Public Health* **71**, 1373–7.

Clark, C. A. M. (1979). *Relation of economic and demographic factors to household decisions regarding education of children in Guatemala.* Ph.D. dissertation. Ann Arbor: University Microfilms.

Cohen, J. (1977). *Statistical Power Analyses for the Behavioral Sciences.* New York: Academic Press.

De Jong, G. F., Root, B. D., Gardner, R. W., Fawcett, J. T. & Abad, R. G. (1986). Migration intentions and behavior: decision making in a rural Philippine province. *Population and Environment: Behavioral and Social Issues* **8**, 41–61.

Engle, P. L., Irwin, M., Klein, R. E., Yarbrough, C. & Townsend, J. W. (1979).

Nutritional and mental development in children. In *Nutrition: Pre- and Postnatal Influences* (ed. M. Winick), pp. 291–306. New York: Plenum.

Engle, P. L., Yarbrough, C. & Klein, R. E. (1983). Sex differences in the effects of nutrition and social environment on mental development in rural Guatemala. In *Women's Issues in Third World Poverty* (ed. M. Buvinic, M. A. Lycette & W. P. McGreevey), pp. 198–215. Baltimore, Maryland: Johns Hopkins.

Fawcett, J. T. (1986). Migration psychology: new behavioral models. *Population and Environment: Behavioral and Social Issues* **8**, 5–14.

Findley, S. (1977). *Planning for Internal Migration: a Review of Issues and Policies in Developing Countries*. Washington, D.C.: U.S. Bureau of the Census.

Fuller, T. D., Lightfoot, P. & Kamnuansilpa, P. (1986). Mobility plans and mobility behavior: convergences and divergences in Thailand. *Population and Environment: Behavioral Social Issues* **8**, 15–40.

Gardner, R. W., De Jong, G. F., Arnold, F. & Cariño, B. V. (1986). The best-laid schemes: an analysis of discrepancies between migration intentions and behavior. *Population and Environment: Behavioral and Social Issues* **8**, 63–77.

Goldstein, S. (1978). Migration and fertility in Thailand, 1960–1970. *Canadian Studies in Population* **5**, 167–80.

Illsley, R., Finlayson, A. & Thompson, B. (1963). The motivation and characteristics of internal migrants: a socio-medical study of young migrants in Scotland. *Millbank Memorial Fund Quarterly* **41**, 115–44; 217–48.

Irwin, M., Klein, R. E., Townsend, J. W., Owens, W., Engle, P. L., Lechtig, A., Martorell R., Yarbrough, C., Lasky, R. E. & Delgado, H. L. (1978). The effects of food supplementation on cognitive development and behavior among rural Guatemalan children. In *Behavioral Effects of Energy and Protein Deficits* (ed. J. Brozek), pp. 415–27. Washington, D.C.: NIH Publication No. 79–1906.

Johnston, F. J., Low, S. M., de Baessa, Y. & MacVean, R. B. (1985). Growth status of disadvantaged urban Guatemalan children of a resettled community. *Amerian Journal of Physical Anthropology* **68**, 215–24.

Kaplan, B. (1954). Environment and human plasticity. *American Anthropologist* **56**, 780–99.

Lasker, G. W. (1952). Environmental growth factors and selective migration. *Human Biology* **24**, 262–89.

Lasker, G. W. (1954). The question of physical selection of Mexican migrants to the United States of America. *Human Biology* **26**, 52–8.

Lasker, G. W. & Evans, F. G. (1961). Age, environment and migration: further anthropometric findings on migrant and non-migrant Mexicans. *American Journal of Physical Anthropology* **19**, 203–11.

Macbeth, H. M. (1984). The study of biological selectivity in migrants. In *Migration and Mobility* (ed. A. J. Boyce), pp. 195–207. London: Taylor & Francis.

Maccoby, E. E., Dowley, E. M. Hagen, J. W. & Degerman, R. (1965). Activity level and intellectual functioning in normal preschool children. *Child Development* **36**, 761–70.

Malina, R. M., Buschang, P. H., Aronson, W. L. & Selby, H. (1982). Childhood growth status of eventual migrants and sedentes in a rural Zapotec community in the valley of Oaxaca, Mexico. *Human Biology* **54**, 709–16.

Malina, R. M., Himes, J. H., Steppick, C. C., Lopez, F. G. & Buschang, P. H. (1981). Growth of rural and urban children in the Valley of Oaxaca, Mexico. *American Journal of Physical Anthropology* **55**, 269–80.

Martorell, R. M., Habicht, J.-P. & Klein, R. E. (1982). Anthropometric indicators of changes in nutritional status in malnourished populations. In *Methodologies for Human Population Studies in Nutrition Related to Health*, NIH Publication No. 82–2462 (ed. B. A. Underwood), pp. 96–110. Washington, D.C.: Government Printing Office.

Martorell, R. M., Klein, R. E. & Delgado, H. (1980). Improved nutrition and its effects on anthropometric indicators of nutritional status. *Nutritional Report International* **21**(2), 219–29.

Pivaral, V. M. (1972). Caracteristicas economicas y socio-culturales de cuatro aldeas ladinas de Guatemala. *Guatemala Indigena* **7**, 5–294.

Pollitt, E. & Gorman, K. (1990). Long-term developmental implications of motor maturation and physical activity in infancy in a nutritionally at risk population. In *Activity, Energy Expenditure and Energy Requirements of Infants and Children* (ed. B. Schürch & N. S. Scrimshaw), pp. 279–298. Switzerland: J.D.E.C.G.

Pollitt, E., Gorman, K., & Engle, P. (1990). Developmental behavioural effects of early supplementary feeding. Paper presented for the Bellagio conference on *The Guatemala Oriente Study*, Bellagio, Italy, July 30 – August 3, 1990.

Population Reports (1983). *Migration, Population Growth, and Development.* Series M, No. 7, pp. 245–287. Baltimore: Population Information Program of the Johns Hopkins University.

Richardson, A. (1974). *British Immigrants and Australia: A Psycho-social Inquiry.* Canberra: Australian National University Press.

Rogers, A. & Williamson, J. C. (1982). Migration, urbanization, and third world development: an overview. *Economic Development and Cultural Change* **30**, 463–82.

Rosenhouse-Persson, S. (1985). *The effects of seasonal migration on fertility: taking into account migration's effects on the proximate determinants of fertility.* Ph.D. dissertation. Ann Arbor: University Microfilms.

Sly, D. F. & Wrigley, J. M. (1986). Migration decision making and migration behaviour in rural Kenya. *Population and Environment: Behavioral and Social Issues* **8**, 78–97.

Teller, C. H. (1973). Access to medical care of migrants in a Honduran city. *Journal of Health and Social Behavior* **14**, 214–26.

Terrell, K. D. (1984). *Labor mobility and earnings: evidence from Guatemala.* Ph.D. dissertation. Ann Arbor: University Microfilms.

Wessen, A. F. (1971). The role of migrant studies in epidemiological research. *Israel Journal of Medical Science* **7**, 1584–1591.

West, D. (1986). *What More Can We Ask For?* Indianapolis, Indiana: Pratt Printing Company.

Wolpert, J. (1965). Behavioral aspects of the decision to migrate. *Papers and Proceedings of the Regional Science Association* **15**, 159–69.

11 Health and nutrition in Mixtec Indians: factors influencing the decision to migrate to urban centres

P. LEFEVRE-WITIER, E. KATZ, C. SERRANO & L. A. VARGAS

In recent decades, Mexico has experienced rapid and drastic demographic, economic and sociocultural changes, which have modified both the conditions of life and the population structure. The Mexican government has made great efforts to improve education, accommodation, hygiene and food quality, and to control tropical diseases (UNDP, 1990). In 1960 the under five mortality rate was 140 per thousand; in 1988 it was 68 per thousand. In 1981 the percentage of one-year-old children with at least one disease was 50; in 1988 it was 75. The adult illiteracy rate fell from 26% in 1970 to 10% in 1985. This process of national development has been slowed by the other changes that are occurring. One such, with far-reaching effects, is the considerable migration to urban from rural areas: the urban population increased from 51% in 1960 to 71% in 1988. This has led to rapid growth of suburban areas in Mexico City, Vera Cruz, Guadalajara and other large towns (indeed, Guadalajara has a population density of 10 286 per square km), where year by year the living conditions have deteriorated, become less acceptable, and less controllable. As a generalisation it seems that the worse the antecedent rural environment of the migrants, the worse the condition of their suburban settlements.

A second major problem is the worsening condition of some rural populations, among them the traditional Indian tribes or communities. Their situation has deteriorated through cultural, linguistic, historical and political isolation much more than through deterioration of their geographical environment. This second problem has led to increased interest by the Mexican government in rural and Indian populations. Several specialised institutions and programmes, to study and help these populations and monitor the changes that are occurring, have been set up. For example in 1945 the National Institute for Indian Studies was founded. In 1973 the Coplamar (coordinacion del plan nacional por zonas deprimidas e marginalizadas), a large programme for the development of poor and remote areas, was initiated, supported by the Institute of Social Security in cooperation with the National Cooperatives. Nevertheless there still remains considerable rural–urban disparity; for example, in 1985–7 access to safe (i.e. piped) water in rural areas was only 53% of that in urban ones, and access to sanitation in rural areas only 17% of that in urban ones.

In 1983 an interdisciplinary research project was initiated jointly by the Paul Sabatier University of Toulouse, France, and the Institute of Anthropological Research of the National Autonomous University in Mexico City. Its object was to examine the relationship between nutrition, health, and urban migration, and so (of direct relevance to the present discussion) to establish the factors that influence a group's decision to continue to live and develop in its traditional territory or to migrate to urban centres, and particularly Mexico City.

Materials and methods

The Mixteca Alta (Mixtec highlands) were chosen for the study. This as a whole is an area of extreme poverty. It forms part of the mountains of western Oaxaca and settlement is at an altitude of 5000–7000 feet (1500–2100 m). The population, approximately 100 000 individuals, belongs to the Macro-mixtec family of the Otomanguean linguistic group (Munch, 1983). There are four administrative districts and two principal climatic areas. In the northern part of three of these districts and all of the fourth, the climate is semi-arid and land erosion is very severe, leading to desertification. According to historical sources, this area was very prosperous until it was devastated in the eighteenth century (Spores, 1984) and in 1630 was reported (by Bernardo Cobo) as being 'among the best lands of the New Spain'. In the southern part of the three districts with the influence of the Pacific Ocean, the climate is less dry and land erosion less severe, there is a rich tropical mountain vegetation, and agriculture has developed (for example, there are small coffee plantations in addition to the traditional farms).

The field work was planned to make a comparison of two small communities in the two zones of greatest ecological contrast: Concepcion Buenavista in the dry eroded Coixtlahuaca district, and San Pedro Yosotato on the moist Pacific edge of the Mixteca Alta. Field work in the latter in 1985–6 provided the data for the present preliminary report. In December 1986 the community of San Pedro Yosotato had a population of 600–900 people, comprising some 150 nuclear families. One third of the population migrates for several months each year to earn wages. Migration is mainly to Mexico City, but also to the town of Tlaxiaco and other parts of the country (Hendricks & Murphy, 1981). In 1985, young men started migrating to the United States to live and work for a few months. The majority of migrants from Yosotato in Mexico City live in Ciudad Nezahualcoyotl, a very poor suburb of the capital.

Field work involved several different methods of data collection. For the study of food resources and habitat usage by Dr Katz, the technique of long-term participant observation was employed. Clinical examination of a large proportion of the Yosotato inhabitants was carried out by Drs Lefevre-Witier, Martinez Maranon, Caire, and Hernandez. Samples of blood and faeces obtained from volunteers were examined partly in the field, and partly in the laboratories of haematology, virology and parasitology in Toulouse and Mexico City.

Factors predisposing to emigration

Health

According to national statistics, in Oaxaca state 32% of the mortality is due to intestinal transmissible diseases, in contrast to 14% for the Mexican nation as a whole. Oaxaca has the highest risk of intestinal contamination and complications. This situation would be expected to occur in the better-watered valleys and coastal areas and to be less likely in the mountain areas. However, our results demonstrate a similar risk in the Mixteca Alta. Table 11.1 compares our serological findings in Yosotato in 1986 with those reported in Oaxaca hospital in 1984–85, and Table 11.2 shows the prevalence of parasites in the stools of Yosotato children in 1986 compared with Tlaxiaco schoolchildren in 1985. *Ascaris* ova were present in the stools of 53% of Yosotato children (more than double the incidence in Tlaxiaco). *Entamoeba coli* were present in 34% and *Endolimax nana* in 22% of the Yosotato children but in no Tlaxiaco schoolchildren. The levels of *Entamoeba histolytica* were quite similar (15% Yosotato children, 22% in Tlaxiaco). For *Entamoeba histolytica*, confirmatory results come from the serological tests (31% positive in Yosotato, 24% in Oaxaca). Not only do the Yosotato children carry a much greater intestinal parasite load, but the serological data indicate higher levels of measles and rubella virus, the presence of *Legionella* (absent in Oaxaca) and a higher incidence of *Plasmodium*. Although levels are generally similar in the other disorders tested for, the very high incidences of salmonella, cytomegalovirus and toxoplasma antibodies give cause for concern.

The combination of our clinical observations with the results of the laboratory investigations shows the extreme liability to infection of the intestinal tract in this population, a situation which may lead to many deaths, especially in the rainy season when medical assistance or evacuation to clinics elsewhere is impossible. In addition to the risk of death,

166   *P. Lefevre-Witier* et al.

Table 11.1. *Comparison of serological findings in Yosotato and Oaxaca hospital children*

Type of infection	1984–5 Oaxaca Hospital			1986 Yosotato		
	Number tested	positive	% positive	Number tested	positive	% positive
Brucella	160	32	20.0	112	16	14.0
Salmonella	149	96	69.9	112	80	71.0
Streptococcus haemolysans	145	26	17.8	112	23	20.0
Legionella	—	—	—	112	13	11.6
Measles virus	160	113	76.8	80	76	95.0
Rubeolavirus	160	98	61.2	80	58	73.0
Cytomegalovirus	160	150	93.7	80	77	96.0
Coronavirus	98	97	99	80	80	100.0
HIV (LAV)	160	0	0	80	0	0
Plasmodium	162[1]	6	3.7	144	21	16.0
Entamoeba histolytica	160[2]	47	23.9	144	45	31.0
	160[1]	24	15.0	144	24	17.0
Toxoplasma	160[1]	78	48.7	144	71	50.0
	160[3]	6	3.7	144	5	3.4

[1] Assay by IIF.
[2] Assay by IHA.
[3] Assay by immune response.

Table 11.2. *Evidence of parasites found in stools*

Parasites	1985 Tlaxiaco schoolchildren			1986 Yosotato children		
	Number tested	positive	% positive	Number tested	positive	% positive
Entamoeba histolytica	112	24	24.5	292	43	15.0
Ascaris	112	23	20.5	292	156	53.0
Trichocephales	112	4	3.5	292	3	1.0
Giardia/Lamblia	112	8	7.0	292	32	11.0
Ancylostomoides	112	0	—	292	5	1.5
Entamoeba coli	112	0	—	292	100	34.0
Endolimax nana	112	0	—	292	64	22.0

parasite loads of this degree enhance the risk of associated diseases and impose a chronic reduction in functional efficiency and work capacity on both the individual and the population. These findings in Yosotato are understandable in view of the living conditions. All families live on steeply sloping hillsides, with unstable and muddy soil. All the houses are built of wood and are very damp. Food is usually prepared on the ground without any hygienic precautions or boiling of water, except in rich families. There are no toilets and no social control of defaecation, most of which occurs around the houses. The resulting soil contamination is a principal source of the high health risk.

So far there has been no real effort to change these conditions. Two projects concerned with public hygiene have been initiated in the community. In 1975 a few latrines were built by some families; in 1983 the elements of a running water system were installed, and the first pipes and taps came into operation in 1986. There is no technical assistance brought to the village from outside, and the transport of any building materials on the local tracks is difficult and costly.

Access to health care is poor. The closest rural dispensary is located in a neighbouring village three miles away. A physician and nurse visit Yosotato but very infrequently. At their clinic, free examinations and medication are provided, but the experience and training of the staff are quite limited. The physician usually comes immediately after leaving medical school to spend two years carrying out the 'social service' required to receive the final medical diploma. His integration with the local population depends on how he accepts working in such a poor area and on his dedication to his profession. The nurse is a young woman from the community who was trained locally by the doctor.

Recognising these limitations, many people of Yosotato seek assistance elsewhere. Some visit the Tlaxiaco dispensary, which is better equipped. If they have enough money, they go to a private doctor at Tlaxiaco or Putla, the main town of the neighbouring district. If they have relatives living in Mexico City, they go there to visit a private doctor, dispensary or hospital. If long courses of treatment are required, they stay in the city until they are cured, a common practice among elderly people whose children have migrated to the city.

Simple improvements to the housing conditions, cooking methods, together with some basic education in hygiene, would reduce the health risk from faecal contamination, but year after year virtually no new initiative has come either from within the community or from outside. But the population of Yosotato is beginning to understand better the lethal risk from intestinal (and other) diseases, they are beginning to understand that prevention is possible, and they know that treatment is available elsewhere.

This they are learning from what they see on their television screens. The decision of many to look elsewhere for improvement of their health is understandable.

Food resources

In the difficult but lush environment of San Pedro Yosotato, the people cultivate small plots and kitchen gardens (0.25–3.0 hectares) and yields are not very high. Traditional subsistence crops (beans, squash, several varieties of maize) as well as other food plants are raised, and there are numerous fruit trees. Gathering of wild plants plays an important part in subsistence, most being weeds collected in the course of agricultural work and consumed as greens. Most families raise pigs and poultry, but only a few have goats, sheep or cows. Hunting and insect gathering occur from time to time. Coffee, introduced at the beginning of the century, is well adapted to the subhumid slopes and is cultivated as a cash crop.

The traditional agriculture and the occurrence of rainy and dry seasons produce a very marked seasonality in the subsistence pattern, with an alternation of times of scarcity and plenty. Maize, beans and squash, sown in May at the beginning of the rainy season, are harvested between October and January depending on the altitude. The minority of families who hold smaller irrigated lands obtain a second crop from them in June, which provides their food at the beginning of the rainy season. Maize is very scarce until the crop matures at the end of the rains, so August and September are called 'months of hunger'. A second period of severe food shortage is February–March. In this period the diet was traditionally limited to peppers, insects and larvae, and edible plants collected in the fields or the forest. In the July–August scarcity period this fall-back diet was supplemented with mushrooms and wild seeds, which are plentiful during the rainy season. There are thus two periods of food shortage each year, and most families are self-sufficient in staples for only three to five months. About 1930 two changes occurred. Up to that time, in the dry season the people of the community used to work temporarily in the neighbouring haciendas, which disappeared in the 1930s. Secondly, by about 1930 coffee production was well established; it brought about a change in the farming calendar and permitted an improvement of the diet, particularly during February–March. Coffee is harvested in the dry season from October to March, so that from its sale eggs and meat, the most precious foodstuffs in the Mixteca Alta, can be purchased for two or three months. A recently introduced cooperative system of coffee production and marketing has consolidated this change in the annual diet.

Thus today the people obtain money during the dry season from coffee production and from their relatives who have migrated, and this allows them to buy staples. If they have enough money, they immediately consume the grain that they buy, and keep their own crops for the rainy season. In the rainy season, even people who have money experience shortages for they find it difficult to bring food from outside, since the mountain tracks become very bad. Even today, in order to cook enough daily tortillas, the housewives mix mushrooms or bananas in with the maize dough; older people recall a similar use of flour from acorns, mango seeds and banana roots. In the rainy season also, the very abundant wild greens often replace beans, mushrooms are cooked in the same way as meat, and the daily chili sauce is made without tomatoes. In the dry season of plenty, several religious and civil festivals are celebrated, with abundant meals. Animals may be killed and eaten. Formerly the Yosotato people used to sell lard and eggs in order to get money. Today they consume more meat, lard and eggs than they produce. Thanks to the income from the coffee, a number of families eat these products regularly in the dry season, but in the rainy season these foods become a real luxury because of the lack of money and difficulty of transportation.

Thanks to its coffee production, this community as a whole is in a better socioeconomic situation than most rural areas in southern Mexico, as is shown for instance in the higher consumption of eggs and meat. The complex traditional use of the habitat, combining polyculture, gathering, and a little animal raising and hunting, allows for a diversity in food resources over and above the diet of maize, beans and chili peppers (Casas *et al.*, 1987). The results of our study at a family level, however, show that cash crop production, besides its benefits, accentuates the economic differences deriving from inequality in land tenure. In 1986 most families harvested about one tonne of coffee per year, and produced staples for 3–6 months; they ate meat once a week and eggs three times a week during the dry season. Some families who had extra income from trade or salaries in the cooperatives produced over 2 t of coffee and enough staples for the whole year; they ate eggs and meat more frequently, and often included festive foods in their daily diet. By contrast some families produced less than 0.5 t of coffee and had to work for the richer ones, and ate eggs and meat less often. Table 11.3 sets out levels of production, income and food consumption, and shows how these variables combine to differentiate three status levels of wealth and health. These differences are more pronounced in the rainy season, since the poorer people cannot store as much food in advance.

In Yosotato, coffee production has for years been beneficial for the community. Food production and supply has improved through the development of a cash economy. It has brought about a solidarity and the

Table 11.3. *Levels of wealth as defined by combining items of production and consumption*

	Status 1	Status 2	Status 3
Coffee production: tonnes of coffee sold to the cooperative market.	2	1.5	0.5
Staple production: comparative production of staple food.	+ + +	+ + +	+
Extra income: comparative importance of extra income such as salaries.	+ + +	+	+
Meat consumption:cannot be evaluated in terms of weight but by frequency (no. of times eaten per week).	1–5	1	0.5
Eggs no. of times eaten per week.	3–7	3	3
Festive food: consumption is frequent and abundant for people of status 1, infrequent for others.	+ + +	+	+

recent establishment of a cooperative is an example of successful local management. It has improved marketing, returns and supplies, for it is able to deal more efficiently with the centralised Instituto Mexicano del Cafe. On the other hand, it has generated socioeconomic heterogeneity among the villagers because of the differences in size of land holding and because some families have to work for others and to give up the traditional collection of edible plants. Thus the nutrition of many families, particularly those of the lowest status, is now linked to the fluctuation of the coffee market and to the diseases of coffee which can ruin a crop. If any important problem affects the coffee harvest and disposal, it is to be expected that possibly half the inhabitants of Yosotato will move to the town, probably to Mexico City.

Discussion

The improvement of the diet as a result of producing and selling coffee offers to some of the Mixtecos a good chance of continuing to exist satisfactorily in this area. However, this improvement is limited by the epidemiological load of intestinal parasites and transmissible diseases. This load not only generates local conditions that increase morbidity and mortality, but it also diminishes productivity, diverting more of the food consumed to the support of the parasites, and it causes permanent malabsorption of nutrients due to local inflammation and other intestinal changes.

Our findings may be summarised in the form of a balance sheet. On the debit side Yosotato has the highest prevalence of intestinal parasites so far observed anywhere in Mexico. The conditions of hygiene are deplorable. The growth of children is delayed. There is no health centre and no permanent medical or nursing staff. The roads are bad, leading to isolation especially at those times when communication is most needed. On the credit side there are the favourable geographical conditions, with the humidity of the Pacific slopes, the rains and the consequent rich vegetation and suitability for cultivation in the area. The people have an excellent knowledge of the local ecosystem, to which the traditional pattern of exploitation is well adapted. The production of coffee for cash provides the key to unlocking the shackles of subsistence level of living. The interaction of these three positive elements leads to the possibility of food availability throughout the year.

Overall, then, this balance is far from negative. There seems no real reason why, given attention to the facts demonstrated in this study, the Mixtec 'people of the clouds' in this region should suffer the same fate as others elsewhere in the Mixteca Alta, where aridity and soil erosion are severe, there is gross undernutrition, and half of the population has already migrated to Mexico City. The ambitious Coplamar project endeavoured to help too many communities, and had to cease part of its activities following the oil crisis of some ten years ago. For a number of years the Mixtecos themselves have supported the idea of a cooperative Mixtecan development project to modernise local production and the economy, in order to escape the 'social privations' to which they and so many other Indian populations are subject. In 1987 such a development project was offered to them by the Oaxaca State Government under the name of Lluvia, Tequio y Alimentos (rain, community work and food). Its goals are to regenerate the soil, to extend irrigation, and so to produce more food using improved seeds and improved technology. This project has every chance of success, provided that it takes into account the factors and their interaction pointed out in this study.

References

Casas, A., Viveros, J. L., Katz, E. & Caballero, J. (1987). Las plantas en la alimentacion mixteca: un aproximacion otnonotanica. *America Indigena* **47**, 317–43.

Hendricks, J. & Murphy, D. A. (1981). From poverty to poverty; the adaptation of young migrant households in Oaxaca. *Mexico Urban Anthropology* **10**, 11–19.

Munch, G. (1983). *Etnologia del Istmo Veracruzano*. National Autonomous University of Mexico.

Spores, R. (1984). In *The Mixtecs in ancient and colonial times*. (ed. J. Norman). Oklahoma University Press.

United Nations Development Programme (1990). *Human Development Report 1990*. Oxford University Press.

12 Urban health and ecology in Bunia, N.E. Zaïre, with special reference to the physical development of children

J. GHESQUIERE, E. NKIAMA, R. WELLENS AND P. DULIEU

Introduction

Bunia is the chief town of the Ituri region, part of the Northeast Zone of Zaïre, situated at 1°35' N and 30°15' E, at an altitude of 1250 m on a hilly, fertile plateau. It has a humid, tropical climate, warmest in April–May and more temperate in July–August. Average humidity is 87%, with a rainfall all through the year of between 1200 and 1400 mm, heaviest in August–September, and drier in December–January. Originally a forest area, it was gradually transformed by slash-and-burn into short grassland. At present, sweet potatoes, cassava, beans and various vegetables and fruits are grown, which make up the main staple foods of the area. The herding Hema provide beef; goat, sheep and pig meat are also consumed. Fresh and smoked fish come mainly from nearby Lake Mobutu (formerly Albert Lake), where fish are abundant and local and industrial fishing is practised. Lately, more meat and fish has been shipped by road to the cities of Kisangani, Beni and Goma, and flown to Kinshasa. These exports, together with gold-panning in the area, have driven up prices, and many town people deriving their main income from wages have a hard time coping with inflation.

The original inhabitants of the region were Lendu farmers and Hema herders. They are now a minority among later immigrants, who come from three distinct ethnolinguistic groups: Bantu (Bira, Nande and others) from the south, Sudanese (Lugbara, Lendu, Ngiti and Lese) from the northwest, and Nilotics (Hema, Alur) from the northeast (see Vansina, 1965.) They derive their income mainly from trade, since Bunia is a traditional trading centre between Kisangani (on the Zaïre river), the south of Sudan, and Uganda. Others are employed in the administration, or provide various services.

It would, however, be misleading to relate the food intake of a Bunia family directly to the father's income or profession. Social stratification is usually based on the income of the father, head of the family. In Central Africa, the father's salary does not necessarily determine the overall living standard, still less that part of the family income reserved for nutrition. More often than not, the father's monthly salary will not make ends meet, and the family has to supplement its income. Whatever its level of instruction or education, each family has to indulge in alternative activities with great diligence, selling peanuts, fresh drinks even alcohol, or providing unofficial or less acceptable services. Hence, socioeconomic distinctions are difficult to establish: as in other parts of the continent, even South Africa, there is as yet no clearly defined system allowing comparison of socioeconomic status (Katzenellenbogen *et al.*, 1988). Even more, the use of such a system in assessing the nutritional status of a family may be very misleading, as the food intake has little or nothing to do with the monthly salary of the father. Instead, it is derived mainly from profits made by the mother and other members of the family. One wonders indeed how many of these enlarged families manage to eat decently, showing no signs of malnutrition. In fact, many practise subsistence gardening around their houses, or cultivate small fields on the outskirts of town. Usually, they grow beans, sweet potatoes, peanuts, soya, cassava and vegetables such as spinach, tomatoes and eggplant. In addition chicken, ducks, rabbits, pigeons and even goats are kept around the house. Father or mother may alternate his or her commercial activities with agriculture; sometimes a jobbing workman will be hired for the more strenuous tasks on the field. The harvest will be stowed at regular intervals, and surplus can be sold to provide money for buying clothes and other products, or to be saved or reinvested in the small trade.

Two meals usually are taken per day, with some snacks in between; the main meal is consumed in the evening, while the morning meal often consists of left-overs from the day before. A large number of children come to school in the morning having had no breakfast, but only some tea.

Health care is provided in different ways. There is the General Hospital, staffed by at least four general practitioners which has to sustain itself largely by its own returns. As a result, medication is in short supply, and so are up-to-date equipment and maintenance. Supposedly free, medicine is expensive and hard to come by. Although there are only two officially qualified chemists, 30 to 40 shops sell medicine, with hardly any control.

Four other self-reliant services are provided: a rather well-kept Medical Laboratory; an extended Vaccination Programme (P.E.V.) that provides free vaccinations, but with insufficient means to reach the whole population; a Centre for Physically Handicapped, caring and providing braces mostly for poliomyelitis victims; and four Health Centres spread over the residential areas, where pre- and post-natal clinics are organised by nurses, supervised by medical doctors from the General Hospital. In conjunction, the local radio station (CANDIP), an offshoot of the Institut Supérieur Pédagogique (I.S.P.) helps to spread information. Even so, attendance at these clinics remains low. Besides this, there is a proliferation of clinics and dispensaries organised in a private way by the physicians, medical assistants or nurses.

Traditional health care is also provided by numerous healers of various ability, who have their own association. Local bill-postings indicate where to find them. They will treat almost any illness, with a preference for patients with good prognosis. Even so, their treatment tends to assuage more than heal, and their advice is mixed with hoodoo, spells, and fortune-telling about projects, weddings and fertility. A surprising number of these healers come from neighbouring Uganda.

Of the Public Health Service during colonial days, only a skeleton remains. Now and then an inspection of housing plots and toilets takes place, but regular garbage collecting is not organised in a systematic way: each family digs its own cesspool, and sometimes garbage is just thrown out on the street.

Running water is provided by the government's 'Régideso', but more efficiently by a missionary backed project (Projet Ngongo). Not all town wards are served. The more outlying neighbourhoods rely on wells, creeks, or the Nyamukau and Ngezi rivers, which collect the rainwaters of the townships. This causes the spread of bilharzia (schistosomiasis) and amoebic dysentery. Malaria, typhoid fever and measles are common; leprosy and plague are endemic.

Bunia is not unlike many of the fast-growing urban agglomertions in Central Africa. In 1986, 59 646 inhabitants were counted; in 1988 there were 64 320. Over the past 10 years, the rate of population increase was close to 4% per year.

With the help of the I.S.P. students, we conducted surveys on two of the most densely populated wards of the town: Sukisa, the oldest part, and Rwambuzi, a slightly better-off and newer district (Nkiama *et al.*, 1988, 1989; Dulieu *et al.*, 1990).

Fertility

We may define the global fertility rate (G.F.R.) by the number of live births per 1000 women at the age of reproduction, on average between 15 and 49 years of age. One can also relate the fertility rate to age, by observing the number of live births within a given age cohort of women, and relating this

Table 12.1. *Global fertility rate (G.F.R. per thousand) from 1977 to 1988, for the wards of Rwambuzi and Sukisa in the town of Bunia*

Year	G.F.R.	
	Rwambuzi	Sukisa
1977	232	—
1978	208	—
1979	230	225
1980	214	248
1981	236	238
1982	260	199
1983	273	309
1984	233	284
1985	268	254
1986	307	290
1987	272	290
1988	199	—
mean	246	256
s.d.	33	37
standard error	10	14

to the female population of this age group. Table 12.1 gives the G.F.R. for the two wards over the years from 1977 to 1988. On average, this G.F.R. is 246 per thousand for Rwambuzi, and similar for Sukisa: 256 per thousand (Nkiama *et al.*, 1988, 1989; Dulieu *et al.*, 1990).

Abortions, stillbirths and deaths

The number of live births per 100 women is given in Table 12.2. By abortions, we mean an interruption of pregnancy, voluntary or not, before the sixth month, expressed as a percentage. This number is given in Table 12.2, as is the number of abortions per 1000 live births. Finally, the percentage of stillborn babies, and child deaths registered over a period of 0–5 years, have been calculated for all women between the ages of 15 and 54. They do not vary much, although one may note a lower abortion rate for Rwambuzi. However, these figures are far below those registered on continental Europe.

Table 12.2. *Comparison of live births, abortions, stillbirths and infant deaths per 100 mothers, and number of abortions per 1000 live births, calculated for all mothers 15 to 54 years old*

Ward	Live births	Abortions	Stillbirths (%)	Deaths (%)	Number of abortions per 1000 live births
Sukisa	325	23	6	49	71
Rwambuzi	394	21	8	44	52

Table 12.3. *Differences in mortality rates for Rwambuzi and Sukisa*

	I.M.R. (2)+(3)	Late foetal mortality (1)	Early neonatal mortality (2)	Perinatal mortality (1)+(2)	Post-neonatal mortality (3)	Mortality at pre-school (4)
Rwambuzi						
Mean	73	21	40	61	33	41
St. error	13.4	5.3	10.4	10.9	7.6	17
Sukisa						
Mean	120	19	43	62	78	39
St. error	20.4	7.6	11.0	13.2	13.2	12.1

Infant mortality

The number of children who died before the age of one year (infant mortality rate or I.M.R., (2+3) in Table 12.3) was calculated per 1000 live births for one year of reference. This mortality rate can be split into:

(1) late foetal mortality (after the sixth month), in which we included those who died during labour;
(2) early neonatal mortality, the babies who died within the first 7 days after birth;
post-neonatal mortality, divided into:
(3) those who died between the 7th and 365th day after birth;
(4) those children who died between one and five years of age, or pre-school mortality.

The perinatal mortality includes the late foetal mortality, and the mortality during labour and during the first week after birth (1 + 2). Averages for the years of observation are given in Table 12.3 for both Rwambuzi and

Sukisa. Infant mortality is lower for Rwambuzi. The breakdowns of late foetal mortality and early neonatal mortality (per thousand) are similar for both wards: means ($\pm$s.e.) are 61 ± 10.9 for Rwambuzi and 62 ± 13.2 for Sukisa. Such high mortality is usually linked to poor labour conditions. For Rwambuzi, however, the post-neonatal mortality of 74 per thousand $(33+41)$ is lower than for Sukisa (117 per thousand $(78+39)$). This post-neonatal mortality is generally attributed to diarrhoea, respiratory infections, tuberculosis and malaria, but protein energy malnutrition (P.E.M.) may also play a role (Masse, 1980). Finally the pre-school mortality, often related to P.E.M. and to infectious and parasitic diseases, is similar for both wards: Rwambuzi 41 and Sukisa 39 per thousand.

Infant mortality is often considered a sensitive indicator of the health status of a population. Bearing this in mind, the values per thousand for Rwambuzi (73) and for Sukisa (120) can be compared with those reported for other parts of the world: 130 for Haïti, 129 for Bolivia, and more than 130 for Central Africa as a whole. Gentilini *et al.* (1986) have estimated a rate of 129 per thousand for all low-income countries. By contrast, this figure stands at 10 or lower for West European countries.

Dupin (1969) uses the ratio of infant mortality to pre-school mortality as an indicator of nutritional adequacy. In a properly fed population, this ratio always stands above 10 and may even reach 20. For the Bunia sample, the ratio is very low: only 1.8 for Rwambuzi (73:41), and 3.1 for Sukisa (120:39) (Dulieu *et al.*, 1990).

Anthropometry

Samples of infants between 0 and 60 months from both wards were taken at random. Weight, height (recumbent or standing), upper arm circumference and triceps skinfold were measured according to WHO specifications (Vuylsteke *et al.*, 1984).

The results are illustrated in Fig. 12.1, comparing height and weight of the Sukisa and Rwambuzi infants from Bunia with infants of the Rega people from Kalima more to the south, in Eastern Zaïre (Van Loon, 1987*a*), and with the norms for European children by Gentilini *et al.* (1986). Table 12.4 compares the weight of the Sukisa and Rwambuzi children with those European standards. This shows that 77% of the children from Sukisa do not reach the 100% standard and that nearly 60% do not even reach the 90% standard for European children; in Rwambuzi, more than 80% weigh less than the 100% European standard and more than 60% of them do not reach the 90% standard.

The use of European standards to assess anthropometric measurements obtained from African people living in vastly different conditions has been

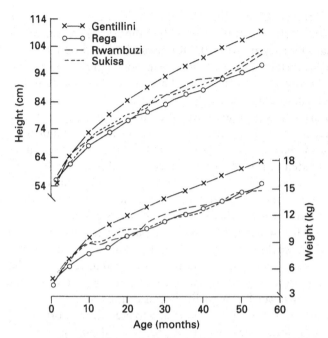

Fig. 12.1 Height and weight of children 0–60 months old from Bunia (Rwambuzi and Sukisa wards) compared with those from Rega (Van Loon, 1987) and references by Gentilini *et al.* (1982).

strongly criticized (Van Loon, 1987*a*). We therefore thought it useful to compare our results with a sample of well-nourished African children with no apparent malnutrition, who also live in the East of Zaïre: the Rega from Kalima (Van Loon, 1987*b*). As can be seen from Table 12.5, the proportion of Bunia children who do not reach the 90% local standard of the Rega is, more realistically, only 16.3% for Sukisa and 22.5% for Rwambuzi. This is also obvious in Fig. 12.1.

Stature or body length is a more stable indicator, less influenced by present malnutrition, although long-term protein deficiency may lead to 'stunting' (Jelliffe & Jelliffe, 1969; Spurr, 1987). Only 4.5% of the Sukisa children and 1.9% of the Rwambuzi sample do not reach the 90% Rega standard.

Upper arm circumference is widely used to assess nutritional status (Jelliffe, 1966). Fig. 12.2 compares the triceps skinfold and the upper arm circumference of the Bunia infants with the same measurements on Rega and the standards of Gentilini *et al.* (1986). Of the Sukisa children, 1.2% had an arm circumference below the 90% Rega standard; for Rwambuzi, this figure was 1.4%. As for the triceps skinfold, an indicator of fat tissues

Table 12.4. *Rwambuzi and Sukisa children classified for weight deficit according to Gentilini's cutoff point*

	Weight related to standard					
	>100%	90–100%	80–90%	70–80%	<70%	Total
Rwambuzi (%)	17.8	21.0	22.9	19.15	19.15	100
Sukisa (%)	22.5	18.0	29.2	21.3	9.0	100

Table 12.5. *Rwambuzi and Sukisa children classified for weight deficit by applying Jelliffe's cutoff point to norms for Rega children*

	Weight related to standard					
	>100%	90–100%	80–90%	70–80%	<70%	Total
Rwambuzi (%)	59.8	17.75	14.95	5.6	1.9	100
Sukisa (%)	56.7	27.0	12.4	2.2	1.7	100

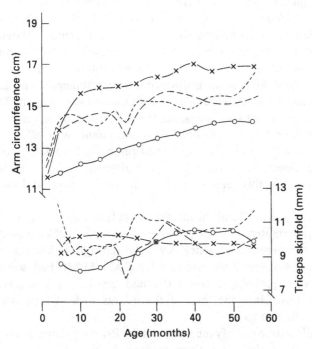

Fig. 12.2 Upper arm circumference and triceps skinfolds of children 0–5 years old. (Symbols as for Fig. 12.1.)

reserves and highly influenced by caloric malnutrition ('thwarting'), 24.4% of the Sukisa children and 31.3% of those from Rwambuzi had skinfolds less than 90% of the Rega standard.

Using Jelliffe's cutoff point based on the Rega standards, we can calculate that 15% of the Rwambuzi children show first-degree malnutrition, as do 12.4% of the Sukisa children; 5.6% show second-degree malnutrition against 2.2% for Sukisa; and 1.9% third-degree against 1.7% for Sukisa. If we use Gentilini's standards, we obtain 22.9%, 19.15% and 19.15%. This would mean that, together 61% of the Rwambuzi children are to be classified as malnourished, which seems surprising when we look at Fig. 12.2. To us, it seems more appropriate to conclude that Western standards are not adequate for describing the African reality.

Clinical signs of malnutrition

Two hundred and forty-one children from Rwambuzi and 178 from Sukisa, aged between 0 and 60 months, were examined for clinical signs of malnutrition. For Rwambuzi, 111 out of 241 infants, i.e. 46.1%, presented at least one clinical sign of malnutrition due to protein or caloric deficiency or due to pathological causes. Among them, 53 (22.0%) presented one, 28 (11.6%) two, 17 (7.1%) three, 8 (3.3%) four, and 5 (2.1%) five clinical symptoms. Of the total of 241 infants, four only (1.7%) presented isolated caloric deficiency of the marasmus type; 48 (19.9%) showed severe protein deficiency or kwashiorkor with distinct hair depigmentation and sometimes uncurling, oedema and muscular thwarting or hematomegaly; and 22 infants (9.1%) showed marasmus associated with kwashiorkor. Again, 22 (9.1%) had a palpable spleen, indicating the likelihood of infectious parasitosis. Finally, no less than 49 (20.3%) of the infants were affected with skin disease, mostly scabies; 30 (12.5%) had gumboils characteristic of vitamin B or C deficiency; and 17 infants (7.1%) had dental caries. Goitre is practically unknown in this region, close to the fishing zone of Lake Mobutu.

As for the Sukisa ward, 22% of the infants had at least one symptom of caloric or protein malnutrition or of pathological disease, 2.2% suffered from aggregate (total) energy deficiency, 3.9% showed signs of kwashiorkor-type protein deficiency, 2.8% presented P.E.M., 11.8% had some symptom of general pathology, and 1.1% had thyroid hypertrophy. Including goitre, this brings the total of the infants with at least one symptom of malnutrition to 10%.

These results allow us to classify the infants of the Rwambuzi and Sukisa wards in Bunia among the populations moderately 'at risk', who will benefit from nutritional and sanitary education of the mothers. This seems all the more urgent as the city wards are in constant demographic extension.

Physical development and fitness of school-age children

The next stage where children can easily be reached for assessment is when they go to school. School attendance is very high in Zaïre, and this provides access to large numbers of growing boys and girls, in town as well as in the villages. The following is based on measurements taken by Nkiama in Bunia on 2154 schoolchildren 6–20 years old (1117 boys and 1037 girls) between 1986 and 1988, and on children from villages in the Ituri forest around Mambasa and Nduye (Ghesquiere *et al.*, 1989) to the west of Bunia, one of the areas from which people migrate to Bunia. The Lese of these villages speak a Sudanese language, representing only one of the three ethnolinguistic groups of which the population of Bunia is made up. Still, we thought it useful to compare data on these children with those from Bunia. We should note that, in the sizeable sample of Bunia children, Nkiama (1985) could find no significant differences in physical development between children from the three main ethnolinguistic origins.

The fiftieth percentile (P 50) for stature and weight of the Bunia and Lese over a background of norms (P 5 to P 95 with P 50 in between) for Afro-American children (Frisancho, 1990) is presented in Fig. 12.3 for boys and Fig. 12.4 for girls. Obviously the Afro-Americans are taller and heavier than the Ituri children. The Bunia boys show a late growth spurt, but barely cross the fifth percentile score for the Afro-Americans by the age of 17. The Lese boys start out close to their cousins from Bunia, but gradually fall back. At the age of 17 years, they lag behind, especially in stature.

At the age of six, the girls have heights and weights close to those of the Afro-Americans, but fall back with advancing age. Their later growth spurt is in agreement with their late menarche: Nkiama *et al.* (1986) studied the age of menarche among a sample of 990 Bunia schoolgirls by a 'status quo' method and found a mean age of 15.7 years, with a median of 15. The growth spurt is minimal in the Lese village girls, while the girls from Bunia nearly catch up with the Afro-Americans by the age of 17. Differences between Bunia and Lese are more pronounced in stature than in weight. Figures 12.3 and 12.4 are illustrations of what Greulich (1951) called the thwarting effects of adverse conditions: we may accept, first, that general hygiene and nutrition is better among Afro-Americans than in the Ituri region, where a further difference can be observed between town children and village children, and second, the notion that adverse conditions will affect the physical development of boys, and only to a lesser extent that of

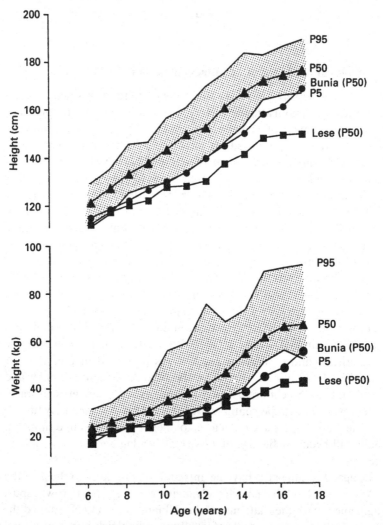

Fig. 12.3 Stature and weight of boys 6–17 years old from Bunia town, and from Lese villages in the forest, compared with norms for Afro-Americans as reported by Frisancho (1990). (P95, P50 and P5 are the 95th, 50th and 5th percentiles, respectively.)

girls. By the age of 17, the Bunia boys barely reach the fifth percentile of the Afro-Americans for stature as well as weight, while the Lese villagers are still way behind. The Bunia girls by contrast catch up to nearly the fiftieth percentile of the Afro-Americans of the same age. Although still behind at the age of 17, the Lese village girls are much closer to the Afro-Americans, their weight reaching the fiftieth percentile of the latter's norm.

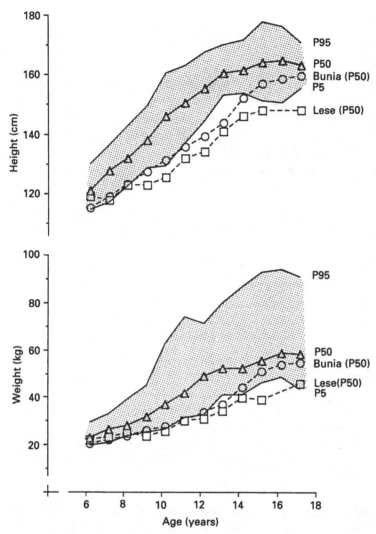

Fig. 12.4 Stature and weight of girls 6–17 years old. (Abbreviations as for Fig. 12.3.)

The ratio of weight to height is often used to give a measure of robustness; Figs 12.5 and 12.6 demonstrate that the Ituri children are relatively lean. To give a better picture, the samples of children were split into age groups 6–11 and 12–17 years. As Waterlow (1990) states, 'Maintenance of low body weight . . . can be achieved by one or both of two ways: to be small in height as a consequence of retarded growth in childhood, and to have a low body weight in relation to height'. It seems the

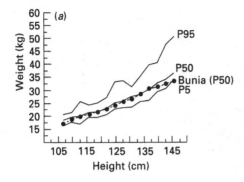

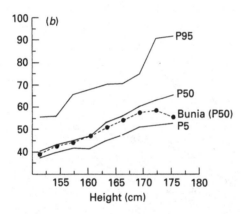

Fig. 12.5 Weight on height for boys aged (*a*) 6–11 and (*b*) 12–17 years, from Bunia town, compared with Frisancho's (1990) percentiles 50 and 5 for Afro-Americans. (Abbreviations as for Fig. 12.3.)

Ituri children adapt by a smaller size but not by a low weight-to-height ratio. This small and light body will also provide thermoregulatory advantages. In a humid environment, sweat does not evaporate and hence loses its cooling effect. Smaller size provides a greater ratio of radiation surface to body mass, and so help to dissipate heat more easily (Hiernaux, 1977), while the lower weight will keep basal metabolic rate (B.M.R.) down.

That food deficiency is not the only adaptive pressure at work seems obvious when we look at Fig. 12.7, giving arm muscle area on stature. The upper arm muscle cross-section area was calculated according to Frisancho (1990), by subtracting the subcutaneous fat layer derived from the triceps skinfold from the total arm cross-section area, derived from the upper arm

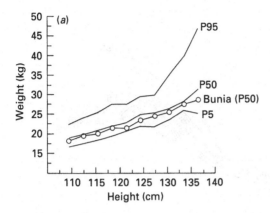

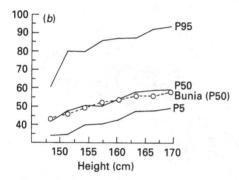

Fig. 12.6 Weight on height for girls age (*a*) 6–10 and (*b*) 11–17 years, from Bunia town. (Abbreviations as for Fig. 12.3.)

circumference. To be fair, we have to relate the arm muscle area of the Ituri children to their smaller body size, as the direct measurements will yield arm muscle areas much to their disadvantage when compared with their taller Afro-American counterparts. The Ituri boys, with no great difference between townspeople and villagers, are leaner than the Afro-Americans but still within the limits of what can be considered 'normal', except for the taller boys. Admittedly, few reach a stature above 170 cm, as the adult male in Bunia stands on average only 168 cm high, while among the Lese the average stature is no more than 165 cm. The Ituri girls, on the other hand, have upper arm muscle areas close to those of their Afro-American counterparts. The Lese village girls present an even greater muscle area than the Afro-Americans over nearly the entire life span considered. At

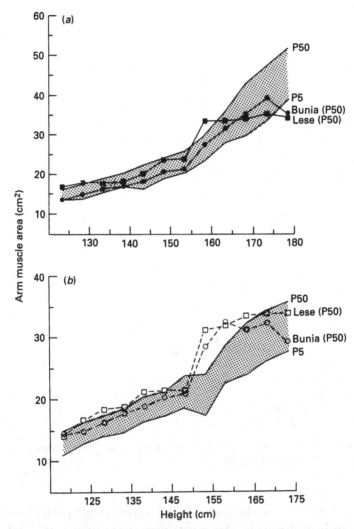

Fig. 12.7 Upper arm muscle area on stature of (*a*) boys and (*b*) girls 6–17 years old, from Bunia town and Lese villages, compared with percentiles 50 to 5 for Afro-Americans of the same size, as reported by Frisancho (1990). (Abbreviations as for Fig. 12.3.)

body sizes of 150–160 cm, when most reach their adult stature, both sets of Ituri girls present larger muscle areas than the Afro-Americans.

The upper arm muscle area has been advocated to assess the body muscle mass (Jelliffe, 1966; Zavaleta & Malina, 1980; Frisancho, 1990). For infants and small children, it turns out to be a good measure (Van Loon, 1987*a*). However, for school-age children it may be misleading, as the arms are not a uniformly trained part of the body in all parts of the world. In some populations or cultures, arms and hands are mainly used as instruments of precision, and less or little for physical 'work', although boys may compensate for this with sports activities. Hence arm muscles get little or no physical training, especially among girls. In other cultures the arms are still used for physical work, to lift and carry burdens, and may get considerable conditioning by activities of daily life (A.D.L.). It may indeed be that the Afro-American children, principally the girls, subjected to a Western urbanised and mechanised way of life, do not use their arms much as instruments of force and hence give them little physical training. In the Ituri, especially in the forest villages, it is quite different: children, principally the girls, will help with the household chores, pound the cassava and millet, carry water and firewood or hoe the field, all A.D.L. tasks that may provide considerable training.

For children from school age on, calf muscle area seems a more appropriate standard, for two reasons:

(1) all children walk on their legs, so the leg muscles, carrying the body mass, are a better, or on average the best trained part of the body (Asmussen, 1968); and

(2) within a given population, there is a high correlation between leg muscles and maximum oxygen uptake ($V_{O_2 max}$.) in children (Cotes & Davies, 1969; Davies *et al.*, 1971; Davies, 1972).

Admittedly, these authors use lean leg volume, and not calf circumference, as a measure of leg muscle mass to predict $V_{O_2 max}$. Studies in our own laboratory on 25 boys aged from 17 to 21 and active in sports show that not much is lost by replacing leg volume, as proposed by Jones & Pearson (1969), by calf circumference: although the correlation coefficient of 0.88 for the relation of $V_{O_2 max}$ to leg volume ($p = 0.01$) came down to 0.72 for $V_{O_2 max}$ on calf circumference, it remained highly significant ($p = 0.01$) (Bosmans, 1982; Van Herp, 1981). Davies *et al.* (1971) and Davies (1972) confirm that, in children, leg volume explains 80% of the $V_{O_2 max}$ obtained during a cycle ergometer test. In the admittedly small sample in our laboratory, the correlation between leg volume according to Jones & Pearson (1969) and calf circumference, both adjusted for stature, was high ($r = 0.78$; $p = 0.01$).

So far we have been comparing our Ituri schoolchildren with the Afro-Americans reported by Frisancho (1990). Unfortunately, Frisancho does not report the calf circumference of these Afro-American children, so we have to use data on Belgian children, for which data are available from an extensive growth study by the staff of our Institute (Ostyn *et al.*, 1980; Simons *et al.*, 1990). Fig. 12.8 compares calf circumference, relative to stature of children from the Lese villages around Nduye, with data from

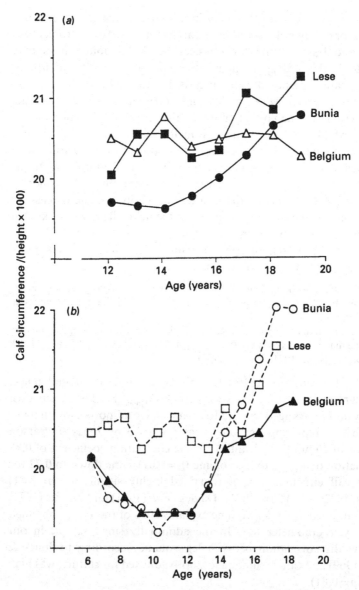

Fig. 12.8 Calf circumference on stature × 100, for (*a*) boys aged 12–19 years and (*b*) for girls aged 6–18 years, compared with data from Belgian children of the same age (Ostyn *et al.*, 1980; Simons *et al.*, 1990).

children from the town of Bunia, and from Belgium. Calf circumference is not equal to calf muscle cross-section area, but the difference is not so important, as the skinfold over the calf tends to be very thin in African children (Eveleth & Tanner, 1976) so calf circumference can be taken to reflect, roughly, the calf muscle circumference.

For their size, the calf circumferences of the Lese boys and Belgian boys looks very similar, while the Bunia boys in our sample have smaller calf circumferences at a younger age, but catch up with advancing age. Upon reaching adulthood at 19 years of age, both Ituri groups have wider calf circumference than the Belgians. The Lese girls apparently have the best developed calf muscles for their body size, while the girls from Bunia have almost identical calf muscles to the Belgian sample, up to the age of 14 years. From 15 years onwards, both samples from the Ituri have wider calf circumferences than the Belgians. Again, the apparently better developed legs of the Ituri girls, especially the Lese, can be the result of conditioning by A.D.L., notwithstanding the poorer health and nutrition to which they are subjected. On the other hand, training and conditioning will have no effect if nutrition is marginal. There have to be sufficient foodstuffs available to the individual to build up his or her body and increase muscle mass. If this is not so, then food will be used to fulfil the energy requirements first, and the body will easily become thwarted. This is obviously not so in the Ituri samples, least of all in the Ituri girls. Therefore, this observation would indicate that nutrition in the Ituri is not marginal. Food is available, although not so plentiful and in such great variety as in Western countries. The slower growth, delayed maturation, and smaller ultimate stature of the Ituri children may reflect adaptation to local environment, of which nutrition, although important, is not the only factor.

Conclusion

The general health and nutrition in a town like Bunia is only fair to marginal, reflected by a high infant mortality and high pre-school death rate. The children who reach school age are significantly smaller than Afro-Americans of the same age. However, the evidence of caloric malnutrition among these children is not conclusive, as their triceps skinfolds are roughly on a par with those of the Afro-Americans. Their upper arm muscle area is similar to if not larger than that of Afro-Americans of the same size. This, to some extent, may reflect conditioning by A.D.L. Furthermore, for the same stature, the calf muscle circumference of the Ituri children is similar to, or larger than, that of their Belgian counterparts. This belies protein malnutrition. So we have to conclude that the Ituri children may be subjected to P.E.M. at younger age, but not so the survivors upon reaching school age. Their small size and light weight can be explained as response to environmental pressure.

Arm muscles will be trained by the daily chores in some societies, while in others the upper limbs are used largely as instruments for precision, so they are not a uniformly trained part of the body in all parts of the world. For that reason the arm muscle mass may result from conditioning as well as from food intake. Legs tend to be more uniformly trained, as all people walk and carry their own weight on their legs. For this reason, we prefer calf muscle area, or simply calf circumference to arm circumference, as a measure of nutritional status for children older than 5 years of age.

References

Asmussen, E. (1968). The neuromuscular system and exercise. In *Exercise Physiology* (ed. H. B. Falls), pp. 3–42. New York & London: Academic Press.
Bosmans, L. (1982). *Beenvolumes en prestatievermogen.* Licenciate thesis, Institute of Physical Education, Catholic University of Leuven.
Cotes, J. E. & Davies, C. T. M. (1969). Factors underlying the capacity for exercise: a study in physiological anthropology. *Proceedings of the Royal Society of Medicine* **62**, 620–4.
Davies, C. T. M. (1972). Maximal aerobic power in relation to body composition in healthy, sedentary adults. *Human biology* **44**(1), 127–39.
Davies, C. T. M., Barnes, C. & Godfrey, S. (1971). Body composition and maximal exercise performance in children. *Human Biology* **44**(3), 195–214.
Dulieu, P., Kahongya, K., Katokolo, L., Nkiama, E., Bazungu, A. & Vitekwisima, L. (1990). Estimation de la mortalité infantile et evaluation de l'état sanitaire et nutritionnel du Quartier Rwambuzi à Bunia (Hau-Zaïre). *Ujuvi, Bunia, (New Pedagogics)* **11**, 1–14. (Institut Supérieur Pédagogique, Departement de diffusion et publication, Candip.)
Dupin, H. (1969). *Les Enquêtes Nutritionelles. Méthode et Interprétation des Résultats.* Paris: Editions Du Centre National de Recherches Scientifiques.
Eveleth, P. B. & Tanner, J. M. (1976). *Worldwide Variation in Human Growth.* Cambridge University Press.
Frisancho, A. R. (1990). *Anthropometric Standards for the Assessment of Growth and Nutritional Status.* Ann Arbor: University of Michigan Press.
Gentilini, M., Duflo, B., Danis, M., Lagardère, B. & Richard-Lenoble, D. (1986). *Médecine Tropicale.* Paris: Flammarion Médecine-Sciences.
Ghesquiere, J., D'Hulst, C. & Nkiama, E. (1989). Fitness and Oxygen uptake of children in the Ituri forest: natural selection or adaptation to the environment? *International Journal of Anthropology* **4**, 1–2; 75–86.
Greulich, W. W. (1951). Growth and development status of Guamean schoolchildren in 1947. *American Journal of Physical Anthropology* **9**, 55–71.
Hiernaux, J. (1977). Long-term biological effects of human migration from the African Savanna to the equatorial forest: a case study of human adaptation to

a hot and wet climate. In *Population Structure and Human Variation* (ed. G. A. Harrison), pp. 187–217. Cambridge University Press.

Jelliffe, D. B. (1966). *The assessment of the nutritional status of the community.* Geneva: W.H.O. Monograph no. 53.

Jelliffe, E. F. P. & Jelliffe, D. B. (1969). The arm circumference as a public health index of protein-calory malnutrition of early childhood. *Journal of Tropical Pediatrics* 15, 177–260.

Jones, P. R. M. & Pearson, J. (1969). Anthropometric determinant of leg fat and muscle plus bone volumes in young male and female adults. *Proceedings of the Physiological Society* 204, 63–6.

Katzenellenbogen, J. M., Joubert, G., Hoffman, M. & Thomas, T. (1988). Mamre community health project – demographic, social and environmental profile of Mamre at baseline. *South African Medical Journal* 74, 328–34.

Masse, L. (1980). *Eléments de Statistiques Sanitaires, Santé et Médecine en Afrique Tropicale.* Paris: Doin.

Nkiama, E. (1985). *Croissance et développement physique des enfants scolarisés à Bunia (Zaïre).* Master's thesis, Institute of Physical Education, Catholic University of Leuven.

Nkiama, E., D'Hulst, C. & Ghesquiere, J. (1986). L'âge moyen des premières règles chez les filles scolarisées de Bunia. *Bulletin et Mémoires de la Société d'Anthropologie de Paris* (ser. 3), 14(1), 27–36.

Nkiama, E., Dulieu, P., Kahongya, K., Katokolo, L. & Vitekwisima, L. (1988). Evaluation de l'état sanitaire du quartier Sukisa à Bunia (Haut-Zaïre). *Ujuvi, Bunia (New Pedagogics)* 9, 47–60. (Institut Supérieur Pédagogique de Bunia, Bureau de Recherches Interdisciplinaires.)

Nkiama, E., Dulieu, P., Kahongya, K. & Katokolo, L. (1989). Evaluation de l'état sanitaitre et nutritionnel du quartier Sukisa (Bunia). *Ujuvi, Bunia (New Pedagogics)* 10, 17–40. (Institut Supérieur Pédagogique, Departement de diffusion et publication, Candip.)

Ostyn, M., Simons, J., Beunen, G., Renson, R. & Van Gerven, D. (1980). *Somatic and Motor Development of Belgian Secondary School Boys. Norms and Standards.* Leuven University Press.

Simons, J., Beunen, G. P., Renson, R., Claessens, A. L. M., Vanreusel, B. & Lefevre, J. A. V. (1990). *Growth and Fitness of Flemish Girls: The Leuven Growth Study.* Sport Science Monograph Series vol. 3. Champaign, Illinois: Human Kinetics Publishers.

Spurr, G. B. (1987). Marginal malnutrition in childhood: implications for adult work capacity and productivity. In *Capacity for Work in the Tropics* (ed. K. J. Collins & D. F. Roberts), pp. 107–40. Cambridge University Press.

Van Herp, W. (1981). *Lichaamsdimensies en prestatievermogen.* Licentiate thesis, Institute of Physical Education, Catholic University of Leuven.

Van Loon, H. (1987). Epidemiology of malnutrition in developing countries. A novel anthropometrical screening strategy. *(a)* Vol. 1, 108 pp.; *(b)* vol. 2, 85 pp. Leuven University Department of Human Biology, Center for Human Genetics.

Vansina, J. (1965). *Introduction à l'Ethnographie du Congo.* Kinshasa: Editions Universitaires du Congo, 229 pp.

Vuylsteke, J., Saverys, V. & Van Loon, H. (1984). *Nutrition en Pays Tropicaux, les Déficiences en Calories et les Malnutritions caloriques et Protéiniques.* Antwerp: Institut de Médecin Tropical d'Anvers, Unité de Nutrition, 159 pp.

Waterlow, J. C. (1990). Mechanisms of adaptation to low energy intakes. In *Diet and Disease in Traditional and Developing Societies* (ed. G. A. Harrison & J. C. Waterlow), pp. 5–23. Cambridge University Press.

Zavaleta, A. N. & Malina, R. M. (1980). Growth, fatness and leanness in Mexican-American children. *American Journal of Clinical Nutrition* **33**, 2008–20.

13 Food for thought; meeting a basic need for low-income urban residents

D. DRAKAKIS-SMITH

Introduction

The objective of this paper is to review the main elements of the food supply systems of the urban poor in the Third World, and how these have evolved and are changing within the broader context of the development process. Geographers are by nature interested in the patterns and interlinkages produced within complex systems of social and environmental relationships and in the processes that shape them. In the context of the urban food supply system (Fig. 13.1), such interest encompasses all levels of the system from production through to consumption. Moreover, within each level, analysis must be focused not just on the commodity itself but also on the economic, social, cultural and political factors that influence the nature of the system.

Although in the biological sciences considerable attention is paid to the nutritional problems and related consequences of inadequate diets, the social sciences have failed to match this with their investigations as to why so many people in the world still have inadequate diets when world food production is steadily increasing. As a result of the patchy coverage of what is clearly a complex system, the relationships between each of the components is imperfectly understood. In turn, this makes policy responses less effective.

Such research deficiencies are difficult to explain when they relate to the most important of basic needs. Low-income families in the Third World usually spend up to 60 or 70% of their income on food (Islam, 1982; Sanyal, 1987). The literature on shelter, in contrast, is enormous, from all shades of the political spectrum.

From the restricted information available, it is possible to identify three main sources of food for the urban poor in the Third World: conventional retail outlets, petty commodity retailing, and subsistence production. In this chapter, I review each of these and attempt to identify the principal trends and related government policies. It must be emphasised at this point that I am not focusing on food solely as an economic commodity, nor on its

193

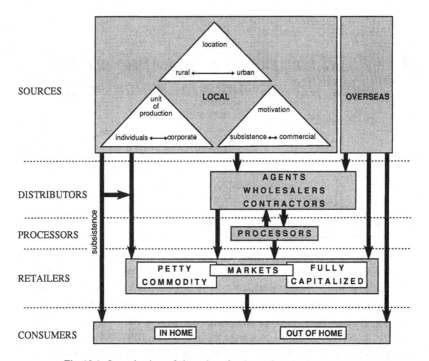

Fig. 13.1 Organisation of the urban food supply system.

acquisition by individuals *per se*, but on the general links between the food supply system and urban society.

Subsistence production

Contrary to popular opinion, there remains considerable evidence of food production within Third World cities, but this does vary considerably within and between countries. It is particularly prevalent in parts of Asia and over much of Africa. The relatively few studies undertaken indicate that it can be very important for the poor. Evers (1983) has shown that 15% of squatter food in Jakarta is self-produced; Sanyal (1987) alleges that for the poorest people in Lusaka subsistence food amounts to one third of total consumption.

Evers (1983) has suggested that the importance of subsistence food varies inversely with income, i.e. that it is most important for the poorest families. However, over much of Africa there is considerable evidence that subsistence food production is a function of the amount of available inputs (land, labour, finance, etc.) so it can also be characteristic of middle-income households (see Gefu, 1990).

In spatial terms, one of the two main sources of subsistence production is urban gardens, particularly in less intensively developed towns and cities. In Harare, Zimbabwe, for example, food crops are grown in about 80% of all domestic gardens (Drakakis-Smith & Kivell, 1990). As cities grow, however, such garden space often diminishes or priorities change along Western lines towards a decorative or an image-enhancing role for the garden. The other principal area for crop production is the urban periphery. This is a complex, ever-changing area (see Hill, 1986), which is very important for the urban poor, offering them opportunities for shelter and fuelwood collection as well as for growing food. Sometimes crop production occurs in small pockets, but usually there are large swathes of illegal or semi-legal production of cereals, and vegetables together with some livestock. The urban periphery can, however, be a very fragile area, not only in the sense that informal land-use may be displaced by the expansion of the built-environment but also because the pressure for basic needs may result in rapid environmental deterioration. The rapid stripping of fuelwood cover, together with increasing crop cultivation or animal husbandry, has frequently led to loss of soil cover and further pressure on remaining areas.

Government reactions to urban cultivation vary but have usually tended to be proscriptive, since its persistence and presence spoils the 'modern' image that so many planners and administrators want for their cities. The consequence has been a tragic burning of crops and a vindictive prosecution of cultivators (Sanyal, 1987). However, the worsening financial position of many Third World countries has forced a *volte-face* during recent years.

Not all subsistence food consumed in the city is grown there. Much still comes into the city from the rural areas. For example, in Harare about one third of the low-income households claim that they receive food from their own land outside the city, and another 20% receive food as gifts from family or friends. As many of these people are also poor, in effect this transfer amounts to a subsidy of the urban poor by their rural counterparts.

It must also be noted that not all food grown in the city is subsistence in nature, i.e. to be consumed by the grower's family. Much of what is grown, particularly by middle-income families or small commercial enterprises, is commercially marketed. However, the important distinction from subsistence production is often overlooked, and all food grown in the city is invariably classified as part of the 'informal sector'. Clearly, the difference is vitally important in relation even to negative government policy. The burning of illegal commercial crops means a loss of income; the burning of subsistence food can mean starvation.

Petty commodity production and retailing

As the poor are denied access to opportunities to grow subsistence food so they must enter the commercial market to purchase it. However, for various reasons (discussed below), conventional retail outlets are often too expensive or too distant and, as with so many other basic needs, the poor must help themselves.

Despite the massive increase in interest over the past two decades in what is usually called the informal sector, research specifically on the production, distribution and retailing of food is quite limited, being confined to relatively well-worn paths such as food hawkers (see Grice, 1989) or urban markets *per se* (see Chandra, 1980). Few investigators have followed Jim Jackson's attempt to look at the full spatial complexity of produce origins, modes of assembly and subsequent wholesaling and retailing (Jackson, 1978, 1979).

What these studies reveal, even within their limitations, is the vital role of the petty commodity or informal sector in meeting the food needs of the poor by making available small amounts at prices the poor can afford (in this context it is the absolute, not the relative, price per standard unit which is important). In Harare, for example, the costly and inconvenient conventional retailing system has been supplemented by a network of fixed-stall hawkers and illegal stores known as tuck-shops, which are scattered throughout the low-income districts of the city (Drakakis-Smith & Kivell, 1990). However, the operations of the petty commodity sector are not solely confined to small-scale activities satisfying the needs of the poor. Jackson (1978) has revealed that many hawkers retail imported goods and, moreover, retail these to all socioeconomic groups within Third World cities (Forbes, 1981).

Observations such as these indicate the complex nature, wide-ranging roles and functionalism of the petty commodity sector, but for many years most of its activities were anathema to the authorities who felt that they spoiled the image of the city and, in addition, were obstructive to the modernisation process. Most myths of this kind were effectively undermined by research in the late 1970s and 1980s, which indicated clearly that the petty commodity sector operated very efficiently and cost-effectively on small profit margins and met a consumer need that was not satisfied by conventional retail outlets. For the poor, in particular, the petty commodity sector is indispensable not only in meeting their material needs but also in providing employment opportunities as well as an important social milieu for migrants to the city (Kaynak, 1981).

The formal or conventional retail sector

It is difficult to draw a sharp distinction between the formal and informal retail sectors: the one tends to grade into the other in a variety of different and highly localised retail types. Shophouses in Southeast Asia constitute an example of this (see Guy, 1986). What should be of concern, however, is the way in which the conventional or formal retail sector has been subject to what MacLeod & McGee (1990) have termed the 'industrialisation' of food supply. This involves the shift to intensive marketing along Western lines and is the consequence of several distinct changes.

First, there has been the impact of indirect changes in lifestyle within developing countries: the desire to follow the Western model in all things. As far as food is concerned, this has induced a change in dietary preferences, which tends to be positively correlated with levels of income but which has usually permeated throughout the socioeconomic scale. Overwhelmingly the change, particularly in urban areas, is away from traditional, indigenous foods towards a Westernised diet, much of which is imported. The second factor relates to the more direct changes in the structural dynamics of food retailing itself. This encompasses the introduction of new technologies, such as refrigeration, which has prolonged the storage, transport and shelf-life of perishable foods; it also covers the evolution of new retailing methods, such as supermarkets and hypermarkets. Although some are pessimistic about the successful penetration of supermarket selling, seeing it confined in impact to the wealthier groups in society (Goldman, 1981), others feel that it has become ubiquitous in most cities in developing countries and constitutes the preferred expenditure mode of all social classes. In Hong Kong, for example, 55% of all food is retailed through supermarkets (MacLeod & McGee, 1990).

The trend towards Westernisation of food consumption and food purchasing is also shown by the massive growth of fast-food outlets. The hawking of cooked food to busy urbanites has always been a feature of Third World cities (see McGee & Yeung, 1977), particularly in Asia, but this has been displaced either by government controlled hawker markets (see Grice (1989) on Singapore) or by the international fast-food chains. In Hong Kong the number of fast-food outlets rose by a staggering 1200% between the mid-1970s and mid-1980s (MacLeod & McGee, 1990).

In all of these changes, which need not always be for the worse, the state has inevitably played a crucial role in facilitating the demise of local and more traditional enterprises, commodities or entrepreneurs in favour of the modern and Westernised. Sometimes such influence is indirect, for example by the introduction of new standards of health and hygiene with which traditional retail enterprises may find it difficult to comply. In particular, the state has sought to intervene to encourage the growth of a food processing industry, much of the output of which is destined for overseas markets. Food processing is by far the leading manufacturing industry in developing countries (Chandra, 1992); and many governments see it as the vanguard of further industrialisation.

Conclusions

The consequences of the changes outlined above have been felt at all levels within the urban food systems of the Third World. At the individual level one of the most worrying effects of the increasing capitalisation or Westernisation of food systems has been the persistence of undernutrition and malnutrition (Sanders & Davies, 1988). This is particularly true for households with erratic incomes (Loewenson, 1988) and/or other important calls on their limited financial resources, such as rent or fuel.

At the urban level, the consequence has been a growing penetration and displacement of subsistence or petty commodity activities by fully capitalised enterprises and commodities not infrequently foreign in origin. At the national level the result has been increased dependency on the advanced capitalist nations and a rising food import bill. Such trends are observable not just in states suffering persistent food deficits but in apparently well-endowed and potentially self-sufficient countries, often with expanding agricultural production. This contradiction occurs because the production of export cash crops has come to dominate the arable areas of many developing countries (Sudarmadji, 1979).

Andrae & Beckman (1985) have vividly documented this process in their book *The Wheat Trap*, which documents the switch from indigenous staples to bread in countries that do not and cannot grow large quantities of wheat. Such trends are unfortunately not new: many former French colonies are still locked into a breakfast habit structured around croissants and baguettes (McGee *et al.*, 1980). Moreover, this is a process that appears to accelerate with the degree or urbanisation, and so is more characteristic in capital cities where the internationalisation of diets and lifestyles is most marked. For example, 75% of Port Moresby's food consumption is imported; this amounts to over one quarter of the total food imports of Papua New Guinea (Harris, 1980).

Some observers allege that one of the additional effects of increasing imports is worsening undernutrition in urban areas because the imported commodities are inferior to local foods (see Thaman, 1982). In more vulnerable situations this has given rise to great concern as imported 'junk' foods become the norm. For example, in the U.S.-dominated Marshall Islands in the Pacific, although 'coconut, papaya and banana trees heavy with fruit sway in the afternoon trade winds . . . kitchens are dominated by canned meats, rice, doughnuts and pancakes heavy with sugar' and what is worse 'local foods are valued less than imported ones' (Anon., 1988).

Furthermore, the increasing preference for and availability of imported foods reduces the incentive for local producers to grow indigenous crops. This in turn induces a downward spiral of dependence as the quality and quantity of traditional food diminishes, causing a further shift towards alternative commodities.

What can or should be done to redeem this situation? Some responses seem self-evident, such as recognition of the importance of subsistence production and traditional foods. However, is it likely that valuable urban land will be released for condoned gardening on the urban periphery? It could be that such usage might be temporary, or that a small charge might be levied, but even this would be preferable to idle land or burnt crops.

It seems evident, too, that a resurgence of demand for traditional foods within the commercial sector must be linked to a programme of consumer re-education into their virtues and practical encouragement to local producers to improve both the quality and the quantity of such crops. Clearly, some degree of subsidy may well be involved, and this alone will be a discouragement to many Third World governments. However, a more positive attitude towards the petty commodity sector could be helpful in this respect, since it is usually very capable of responding to the needs of low-income residents. Whatever reponse is planned, the state must play a central role. It already does this in facilitating the penetration of capitalist enterprises into the urban food market. A determined and sympathetic government could be equally helpful to local producers, retailers and, ultimately, consumers.

However, it must be recognised that there can be no single or uniform solution to the present dilemma. The nature of the problem varies in each locality, although some elements of the process remain consistent. It is also true that the problems of and responses to the evolving urban food system cannot be separated from the broader changes affecting urban society as a whole in the Third World. It would be unrealistic and counterproductive to seek a return to an assumed pre-capitalist idyll.

There are, however, ways of using the process of change itself to achieve new objectives. Many governments in newly independent states are themselves major food purchasers for their school, army, police or civil service canteens. It is in their power to influence directly the nature of local food consumption. Many administrators and planners, however, appear to be appallingly ignorant of the urban food supply system and, even if sympathetic to the nutritional problems of the poor, do not look towards the supply system itself for improvements, falling back instead on appeals

for food aid or attempts to raise incomes by economic growth. Such
monetarist solutions seldom work to the benefit of the poor. Until there is
more research and information available on urban food systems we will not
be fully able to appreciate how particular responses or measures affect the
system as a whole. In short, more thought on food is needed in order to
promote food for thought.

References

Andrae, G. & Beckman, B. (1985). *The Wheat Trap.* London: Zed Books.
Anonymous (1988). War declared on junk food. *Pacific Islands Monthly*, August
    1988.
Bibangambah, J. (1990). 'Macrolevel constraints and the growth of urban informal
    sector. Paper presented at a conference on Small Towns and Rural Develop-
    ment in Africa under Conditions of Stress, Gilleleje, Denmark, 1990.
Chandra, R. (1980). *Food Distribution Systems in the South Pacific.* Canberra:
    Development Studies Centre, A.N.U.
Chandra, R. (1992). *Industrialization and Development.* London: Routledge.
Drakakis-Smith, D. W. & Kivell, P. T. (1990). Food production, retailing and
    consumption patterns in Harare. In *Retailing Environments in Developing
    Countries* (ed. A. Findlay, R. Paddison & J. Dawson), pp. 156–80. London:
    Routledge.
Evers, H.-D. (1983). Households and urban subsistence production. Paper
    presented at a seminar on Third World Urbanization and the Household
    Economy, Universiti Sains Malaysia, Penang, 1983.
Forbes, D. (1981). Petty commodity production and underdevelopment: the case of
    pedlars and trishaw riders in Ujung Pandang, Indonesia. *Progress in Planning*
    16(2), 105–78.
Gefu, J. (1990). Part-time farming as an urban survival strategy: a Nigerian case
    study. Paper presented at a conference on Small Towns and Rural Develop-
    ment in Africa under Conditions of Stress, Gilleleje, Denmark, 1990.
Goldman, A. (1981). Transfer of a retailing technology into the less developed
    countries: the supermarket case. *Journal of Retailing* 57(2), 5–29.
Grice, K. (1989). *Institutionalization of informal sector activities: cooked food
    hawkers in Singapore.* PhD thesis, Department of Geography, University of
    Keele.
Guy, C. (1986). Retail distribution in Third World cities: some research issues.
    Paper presented at IBG Developing Areas Research Group Conference,
    University of Leeds, 1986.
Harris, G. T. (1980). *Replacing Imported Food Supplies in Port Moresby, PNG.*
    Occasional Paper no. 17. Development Studies Centre, A.N.U., Canberra.
Hill, R. D. (1986). Land use change on the urban fringe. *Nature and Resources*
    22(1/2), 24–33.
Islam, N. (1982). Food consumption expenditure patterns in Bangladesh. *Geo
    Journal* 4, 7–14.
Jackson, J. 1978). Trader hierarchies in Third World distribution systems. *Food.
    Shelter and Transport in East Asia and the Pacific* (ed. P. J. Rimmer *et al.*),

pp. 33–62. Monograph HG12. RSPACS. Canberra: Australia National University.

Jackson, J. (1979). Retail development and Third World cities. In *Issues in Malaysian Development* (ed. J. Jackson & M. Rudner), pp. 273–303. Singapore: Heinemann.

Kaynak, E. (1981). Food distribution systems: evolution in Latin America and the Middle East. *Food Policy*, May 1981, pp. 78–90.

Loewenson, R. (1988). Labour insecurity and health: an epidemiological study in Zimbabwe. *Social Science and Medicine* 27(7), 723–31.

McGee, T. G., Ward, R. G. & Drakakis-Smith, D. W. (1980). *Food Distribution in the New Hebrides*. Monograph no. 28, Development, Studies Centre, A.N.U., Canberra.

McGee, T. G. & Yeung, Y-M. (1977). *Hawkers in Southeast Asian Cities*. Ottawa: IDRC.

MacLeod, S. & McGee, T. G. (1990). The last frontier: the emergence of the industrial palate in Hong Kong. In *Economic Growth and Urbanization in the Third World* (ed. D. Drakakis-Smith), pp. 304–35. London: Routledge.

Sanders, D. & Davies, R. (1988). The economy, the health sector and child health in Zimbabwe since independence. *Social Science and Medicine* 27(7), 723–31.

Sanyal, B. (1987). Urban cultivation amidst modernization: how we should interpret it. *Journal of Planning Education and Research* 6, 197–207.

Sudarmadji, S. (1979). Food consumption patterns and the ASEAN food dilemma. *Contemporary Southeast Asia* 1(1), 92–105.

Thaman, R. (1982). Deterioration of food systems, malnutrition and food dependency in the Pacific Islands. *Journal of Food and Nutrition* 39(3), 109–25.

14 *Immunological parameters in northeast Arnhem Land Aborigines: consequences of changing settlement patterns and lifestyles*

G. FLANNERY AND N. WHITE

Introduction

It is understood by all Australians that the health of Aboriginal people has suffered greatly as a result of European settlement and Westernisation. Morbidity and mortality statistics for Aborigines are considerably worse than for non-Aboriginal Australians: infant and perinatal mortality are approximately 2–4 times those of white Australians, with gastrointestinal infection a prime cause; hospital admissions for bacterial and parasitic infection and for respiratory disorders are approximately 5–7 times those of non-Aborigines; life expectancy for Aboriginal men and women is, on average, approximately 17 years less than for non-Aboriginal Australians (Thomson, 1991). While this has been acknowledged by the Australian government and increasing amounts of money and resources are being directed to Aboriginal health problems, 'lifestyle diseases' such as non-insulin-dependent (type 2) diabetes mellitus, cardiovascular disease and alchoholism have become particularly prevalent. Relatively little attention has been paid to the links between changing lifestyles (including settlement patterns, activity and nutrition), immune status, community health and health education, in the formulation of effective health services which take into account differences in cultural beliefs and practices.

There is considerable interest among immunologists and medical practitioners concerning the relationship between lifestyle and immune status: much research has demonstrated that nutritional status and immune responsiveness are strongly correlated (see, for example, Gershwin *et al.*, 1985) and the recent emergence of 'neuroimmunology' clearly links emotional and physical 'stress' with changes in immune parameters (see, for example, Kelley, 1980). Hence it is reasonable to assume that the process of urbanisation (more properly, in the present context, Westernisation) will be reflected in immunological status. It is perhaps surprising, then, that relatively few studies have explored the impact of urbanisation/ Westernisation on immunity and its implications for health.

The present study investigates the biomedical consequences of lifestyle changes among Aboriginal communities living in the Arnhem Land Aboriginal Reserve in the 'top end' of the Northern Territory of Australia. Apart from interest in the general issues outlined above, the project sought not only to describe biomedical changes taking place in Aboriginal communities, but also to help people to understand these changes and to cope with them.

The northeast Arnhem Land human ecology project

Background

The La Trobe University human ecology project began in 1970 and has been described in detail elsewhere (White, 1978, 1985; White *et al.*, 1990). The project began with an investigation of the extent of genetic variation and the genetic relationships among tribes in the Northern Territory, after which the focus of attention has become the people of northeastern Arnhem Land, known collectively as the Yolngu. Occupying an area of about 26 000 square kilometres bounded by coastline and river systems, the Yolngu number approximately 3200, are genetically distinct from neighbouring tribes (but heterogeneous themselves (see White, 1976)) and retain a complex social structure. Unrecorded contacts with pre-British visitors (e.g. Macassans from Indonesia) are evident in Yolngu belief and material culture, but significant changes in lifestyle occurred only in the early part of this century in response to the establishment of mission stations along the Arnhem Land coast. Further dislocation followed the construction of an airstrip on the Gove peninsula during World War II and the commencement of bauxite mining in the late 1960s with the concomitant development of the mining township of Nhulunbuy. The large Aboriginal settlement established at Yirrkala in 1934 has, partly in reponse to these activities, set up bush camps (outstations or homeland centres) of which some 22 now exist within a radius of 180 km of Yirrkala.

Consistent with the Yolngu view of an unchanging spiritual world, the people have retained the key elements of social organisation, kinship and religion, which remain basically unaltered from those recorded by the earliest anthropologists. Important changes have taken place, however, in their material culture, and recently in activity patterns and diet. Thus the people of the eastern area of the Yolngu territory are still quite mobile, but their nomadic existence has been reduced to varying periods spent in the Yirrkala settlement, alternating with (usually briefer) periods occupying one or other of the homeland centres. Lifestyle factors such as time spent hunting and foraging, the availability of indigenous versus processed or Westernised foodstuffs, attainability of other store-bought goods such as tobacco and alcohol, and access to medical facilities such as the hospital at Nhulunbuy and the clinic in Yirrkala, are clearly influenced by the time spent in the main settlement. This in turn is determined by, for instance, the seasonality of game, the availability of vehicles and road accessibility, distance to particular homeland centres and the occurrence of ceremonial gatherings. In addition to (seasonally restricted) road access, many homeland centres now have airstrips, which allow unseasonal and rapid movement of goods and small numbers of people.

The work described in the present study involved a cross-sectional survey of people in the main settlement of Yirrkala and in the homeland centres. In addition, data have been drawn from a longitudinal study in one isolated homeland centre to the west of the region described above: the

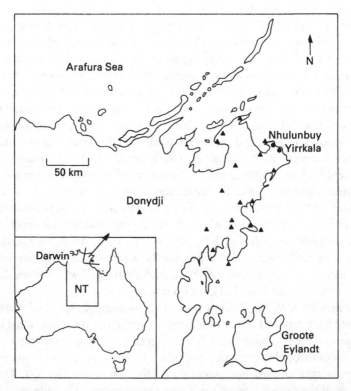

Fig. 14.1 Eastern Arnhem Land, showing the locations of the mining township of Nhulunbuy, the large Aboriginal settlement of Yirrkala, the Yirrkala homeland centres (triangles) and Donydji outstation.

Donydji bush camp (see Fig. 14.1). A permanent camp was established at Donydji in 1968 by the nomadic local families following the construction of an airstrip for mining surveys and the subsequent damage to a sacred site nearby. Residents of Donydji (numbering between 20 and 60 individuals depending upon season) are possibly the most traditionally oriented Aboriginal people in eastern Arnhem Land. People in this and neighbouring communities still rely to a considerable extent on hunting and foraging for their subsistence, although the presence of the airstrip and the recent access to off-road vehicles have resulted in increased European contact and greater dependence upon western foods, particularly white flour and sugar, which are available from a mission township 90 km away. Biomedical data have been collected over a period of 18 years in Donydji by one of us (N.G.W.) assisted by Melbourne-based medical practitioners Dr Kelman Semmens and Dr Alicia Polakiewicz.

The data described below, then, are the results of a large survey conducted in Yirrkala and related homeland centres in 1986, combined with material gathered from the isolated bush camp at Donydji between 1976 and 1990. The detailed data collection made possible by longitudinal studies of the small band at one homeland centre allows analysis of the cross-sectional information from a much larger population and emphasises the need to gather information on diet (including food distribution), subsistence behaviour and lifestyle, if biomedical profiles are to be interpreted meaningfully and health status assessed accurately.

Biomedical data

During a broader biomedical survey of communities in northeast Arnhem Land conducted at the request of local Aboriginal councils, various immunological and haematological parameters were determined. In the 1986 study, with the assistance of trained Aboriginal health workers, a locally based nurse educator and a Melbourne physician, over 400 individuals (aged 10 years and above) were examined. Five homeland centres were visited by aircraft; the sample (almost 50% of the total population) included approximately equal numbers of Yirrkala residents and outstation residents (the latter seen during visits to Yirrkala or in the outstations themselves). In addition, the outstation volunteers included people from coastal and inland camps (see Table 14.1).

Volunteers attended the Yirrkala clinic or attended *ad hoc* 'clinics' set up at outstation airstrips on the days designated for flying visits (see Fig. 14.2). Each participant was subjected to a series of anthropometric measurements (including height, weight, skinfold thicknesses and other body measurements) and received a brief medical check, including heart rate and blood

Table 14.1. *Number of people surveyed according to location, age group and sex*

Location	Age group (years)				
	≤15	16–25	26–35	≥36	Total
Yirrkala settlement					
males	21	29	22	12	84
females	23	31	30	17	101
total	44	60	52	29	185
Homeland centre (inland)					
males	24	14	4	13	55
females	21	14	9	23	67
total	45	28	13	36	122
Homeland centre (coastal)					
males	12	20	7	13	52
females	17	13	5	17	52
total	29	33	12	30	104

pressure determinations. Venous blood was drawn for subsequent immunological and biochemical tests (including haematocrits, total and differential leukocyte counts, blood films and serum for determination of immunoglobulin and specific antibody levels).

Accumulated biomedical data gathered in the Donydji homeland centre include many of those specifically assessed in the broader survey; stools were also collected for determination of parasite infestation. In addition, data include details of activity patterns, diet (including food distribution and preparation), effects of seasonality, foraging strategies and sex differences in those parameters which might influence health profiles.

Results and discussion

General health and biomedical profiles

The people least influenced by Westernisation (in our study population) are those resident in Donydji outstation, and it is these people about whom most is known with respect to parameters such as blood chemistry and the changes in health profiles over time. As described elsewhere (Flannery, 1985; White, 1985; O'Dea *et al.*, 1988) the general health of Donydji residents is good: body mass indices are low, but at the time of the surveys there was no evidence of impaired work capacity; there was little anaemia and no biochemical evidence of dietary deficiency. Red cell folate levels

(*a*)   (*b*)

Fig. 14.2 (*a*) The modern health clinic in the Yirrkala settlement is readily accessible by sealed road and has facilities for patient care and some laboratory testing. (*b*) *Ad hoc* field 'clinic' at the airstrip, Biranybirany homeland centre, where no permanent health workers are resident.

were normal to high and fasting cholesterols generally low. While there is some variation in indicators of nutritional status, this is to a large degree explicable in terms of the non-random distribution of food among these traditionally oriented people (White, 1985). The data are consistent with the essentially traditional diet, but the recent increase in the use of flour and sugar is reflected in the high frequency of dental caries, which were absent when the long-term programme began almost 20 years ago.

Indicators of general health in the large settlement of Yirrkala (and to an extent the residents of its associated homeland centres) contrast markedly with those seen in Donydji. As might be expected of people living a more Westernised lifestyle, body mass indices are generally higher and anaemia (assessed in terms of erythrocyte packed cell volumes (PCV) in the absence of blood chemistry analysis) evident in almost 20% of settlement and homeland centre residents. A study of haemoglobin levels carried out in 1983 (Watson & Tozer, 1986) revealed a frequency of (mostly iron-deficient) anaemia, due in part to hookworm infection, of 11% in the residents of Yirrkala, although the authors suggest that this is likely to be an underestimate. In the present study, analysis by age differences shows that anaemia in males is confined almost exclusively to the young (19 out of 22 males with low PCV were less than 20 years of age); for females the distribution is different, with a higher number of older women affected (14 out of 40 females with low PCV were over 20 years of age and 10 of these were over 40 years). These results are consistent with the effects of menstruation and childbirth, and also with diet and food distribution data obtained from Donydji, in that men, especially men of rank, are nutritionally privileged whereas older women are nutritionally underprivileged (White, 1985). Reviewing the Yirrkala haemoglobin data from 1948 (the American–Australian Scientific Expedition to Arnhem Land) and 1972 (hospital records), White (1985) noted that males, but not females, showed significantly higher values in outstations compared with settlement residents. These differences, together with the general decline in the haemoglobin levels of males living in the settlement reflect sex-specific dietary and lifestyle differences.

Detailed anthropometric and dietary analyses of Yirrkala and homeland centre residents are in progress, but preliminary results (Jones, 1990 and unpublished) indicated that differences in health-related parameters (such as adiposity) are related to differences in diet and hence to locality (settlement versus bush camp). For instance, although total caloric intake is similar in both situations, dietary fat intake is greater in Yirrkala and dietary iron is signifiantly lower, an observation consistent with the haemoglobin values. Further 'location' effects are evident in analysing outstation diets: coastal camps are supplied with fish all year round, for instance, and people there consume the least store-bought food, while inland camps have little access to fish and utilise store food to an extent intermediate between those of Yirrkala and the coastal camps. Clearly, then, locality or most frequented residence is an important determinant of health in these communities.

Health care and its availability also vary according to residence pattern. At Nhulunbuy there is a small but fully equipped hospital, approximately 20 km from Yirrkala. In Yirrkala itself there is a modest but adequate medical clinic, staffed by trained Aboriginal health workers and a nurse and visited weekly or fortnightly by a medical practitioner from Nhulunbuy. Outstation residents, in particular those in distant outstations, are dependent upon Aboriginal health workers living in the communities or moving between them, and occasional flying visits by hospital staff. In some homeland centres the level of health care is minimal and it is often difficult to obtain trained medical help even in an emergency. A number of issues confound this situation, but it is ironic that while a return to bush living is associated with indicators of improved health (see, for example, O'Dea, 1984; O'Dea et al., 1988), for those who fall ill, health care in the homeland centres is frequently inadequate.

Immunological parameters: non-specific measures

Analyses of total and differential leukocyte determinations will not be discussed in the present paper. Suffice it to say that to date we have observed no clear trends of, for instance, leukopenia, nor is there any evidence of T lymphocyte subset disturbances in the limited studies possible so far. Eosinophilia (counts in excess of 10%) is, however, a common finding in all communities surveyed and is probably a response to parasite infestation, although the possibility of parasite-associated allergic reactions is discussed below.

Serum immunoglobulin (Ig) levels were determined by immunodiffusion or by nephelometry, and the mean values for the Yirrkala settlement were compared with those of the pooled homeland centres (Fig. 14.3). Data derived from two other Aboriginal populations (samples from Professor K. O'Dea) and from published findings in two non-Aboriginal populations are included for comparison. It is evident that, with the exception of IgM, where there is considerable overlap, the Aboriginal immunoglobulin levels from all populations are greater than the corresponding values for non-Aboriginal Australians. Elevated immunoglobulins have been reported previously in Aboriginal (Wilkinson *et al.*, 1958) and other non-European (Rawnsley *et al.*, 1956) populations; the values for the Papua New Guinea (PNG) sample shown in Fig. 14.3 are examples of the

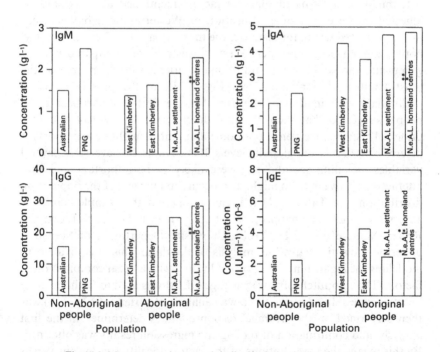

Fig. 14.3 Mean serum immunoglobulin levels in the northeast Arnhem Land (N.e.A.L.) populations with other Aboriginal and non-Aboriginal population values shown for comparison. For 'West Kimberley' and 'East Kimberley' see text. For 'Australian' and for PNG (Papua New Guinea) see Groves *et al.*, 1975; * and ** represent values obtained from Donydji outstation in 1980 and 1985, respectively.

latter. The issue of whether such elevated levels are genetically determined or simply reflect the high incidence of infection, particularly parasitic, in the populations studied has not been satisfactorily resolved. Malarial parasitism is a likely cause of elevated immunoglobulin, in particular IgE, in the PNG sample cited, but the authors (Groves *et al.*, 1975) determined that less than 10% of the IgE was parasite-specific. On the other hand, non-specific elevations in immunoglobulins are a characteristic feature of chronic antigenic stimulation, and it is likely that much of the increase in this case might be parasite-driven, although not being parasite-specific. Parasitic and other infections are prevalent in the Aboriginal populations reported here (see below) and undoubtedly play an important part in determining the levels of serum immunoglobulins. Other factors too, influence immunoglobulin levels: allergic responses, diet (in particular trace elements and, in parasitised individuals, low calorie or protein intake), exercise, stress, and the use of tobacco and alcohol may all contribute.

It might be anticipated that the factors mentioned above would be reflected in the extent of urbanisation or Westernisation, but no clear picture emerged when, for instance, the more urbanised 'West Kimberley' population of Derby in northern Western Australia was compared with the genetically similar but more traditionally oriented desert people of the 'East Kimberley' population (see Fig. 14.3). In the present study, a preliminary examination of the data for the northeast Arnhem Land people suggested that such 'locality' differences were evident and that principal residence (settlement or homeland centre) might be an important determinant of immunoglobulin levels. It is obvious, however, that several variables including sex and age were likely to be important and that interactions between them might also occur, and in view of the population distribution (see Table 14.1) it became apparent that simple statistical analysis was therefore inappropriate. For this reason, Drs Chris Lloyd and Robert Staudte of the La Trobe University Department of Statistics, undertook logistic regression analysis of the data. This is equivalent to modelling the logarithm of the expected values with a linear combination of factors. The computer software package GLIM was used to fit these linear logistic models. In addition, three-way analysis of variance (ANOVA) was then performed on the 'average' response values determined in the first analysis, and confirmation of the logistic regression results was obtained. By this means the contributions of the factors locality (settlement versus outstation), sex and age were assessed independently and in terms of any interactions. Modelling of this sort does not take into account the lifestyle variables associated with locality (such as alcohol, exercise or diet), but observations of behaviour and diet derived from the long term field studies

Table 14.2. *Summary of statistically significant[a] associations between immunoglobulin levels, locality, age and sex*

	Immunoglobulin			
Variable	IgM	IgG	IgA	IgE
Locality				
H = homeland centres	H > S	H > S	H > S	H > S(A)
S = settlement				H < S(C)
Age[b]				
C = children or young adults	A < C	A < C(H)	A > C	A > C(S)
A = older adults				A < C(H)
Sex				
F = female	F > M	No	F < M(H)	F > M(A,H)
M = male		gender		F < M(A,S)
		difference		

[a] Associations based on logistic regression analysis and three-way ANOVA (see text).
[b] C = individuals up to 25 years of age; A = individuals over 25 years of age.

in Donydji and the Yirrkala region can then be utilised in explaining further some of the differences observed.

A summary of the statistically significant associations between immunoglobulin levels and other variables is presented in Table 14.2. For IgM, location, age and sex are all important variables in explaining the distributions, but no interactions are necessary. Thus higher levels of IgM are found in homeland centres, in females and in younger individuals. The sex difference in IgM levels apparent in the West and East Kimberley Aboriginal populations (data not shown) has been reported previously in European populations (see, for example, Zegers *et al.*, 1975) but is not a universal finding. No explanation has been provided, but the immune responses to many infectious agents are greater in females than in males in humans and in experimental animals and may be influenced by hormones. The sex differences might also reflect differences in exposure to, for instance, parasites: females might spend more time in camp, exposed to soil-borne parasites around water sources or to faecal parasites. The age effect seen here has not been reported previously, and although reinforced by the inhibitory effect of tobacco smoking (a predominantly adult pursuit) in this instance possibly reflects higher levels of 'new' or recurrent infections, as IgM constitutes the early antibody response. Certainly, for some infective organisms, children have a higher frequency of infection (see below) consistent with this idea. Similarly, if the elevated IgM levels represent early responses to infection, then the higher levels found in homeland centres might be explicable in terms of a greater likelihood of exposure to parasites in bush camps where, for instance, sanitation and water supplies are a problem.

The IgG data indicate that the effect of location differs across age groups but that there is no evidence of a sex effect on the overall concentration levels or as an interacting factor on the effect of location or age. Once again the homeland centre values are higher (for both sexes) than the values in Yirrkala settlement. No consistent sex differences have been reported for IgG, nor have consistent age effects been seen, although in the present study the homeland centre population shows a higher frequency of young individuals with high levels of IgG (69% and 38% of young with values $>25\,\mathrm{g\,l^{-1}}$ for bush camps and Yirrkala, respectively). This might be explicable in terms of the infection rates among children, although IgG appears only late in primary infections and in secondary immune responses. Another factor that might influence the locality effect is the availability of medical treatment: residents of the main settlement are more likely to be treated for illness, or to be treated soon after its onset, given the access to the Yirrkala clinic and the visits of hospital staff from Nhulunbuy.

For IgA, both logistic regression and analysis of variance indicate that locality, sex and age are important factors, but an age–sex interaction is only strongly indicated in the former analysis. For each demographic group, the homeland centre levels are higher than the corresponding settlement values; again, as for the elevated homeland centre IgM and IgG levels, this is likely to reflect infection rates. Amongst older outstation residents, this effect is most apparent, as is a trend towards higher levels in males. Neither the age effect nor the sex difference have been reported for other populations, but it is interesting to note that the sex difference is also observed in the Kimberley Aboriginal populations (data not shown). Secretory IgA is of major importance in protecting mucous membranes, but elevated serum IgA levels are associated with pulmonary infections such as tuberculosis, as is elevated IgG. Medical examinations of Donydji residents indicate a high frequency of respiratory tract infection, the nature of which is unknown. It is possible that parasitic and other infections are exacerbated in older individuals, especially males, by heavy consumption of tobacco. It might be anticipated that the transitory stimulation of IgA by alcohol would be most apparent in the settlement, which although it is alcohol-free, is in close proximity to the township of Nhulunbuy. The data, however, show no evidence of such an effect.

The IgE data are perhaps the most difficult to interpret. There is no statistically significant location effect nor age or sex effects. The logistic regression analysis suggests, however, that although the mean levels are not affected, the distribution is dependent upon locality and age. While the modal values are the same in the Yirrkala and homeland centre groups (Fig. 14.4), the distributions differ. Adults have a higher average value than children in the settlement, but the reverse is the case in homeland centres. Locality and sex differences in the distribution are also evident in the analysis of variance (see Table 14.2). In the homeland centres fewer individuals exhibit 'extreme' concentrations, so that most individuals cluster around 1000–3000 I.U. ml^{-1}, while in the settlement, a broad distribution is seen, although again the greatest frequencies lie in the region of 1000–3000 I.U. ml^{-1}. Few data are available concerning 'healthy' individuals and IgE, but the values cited for Europeans (see Zegers *et al.*, 1975) suggest that the Aboriginal values reported here for all populations are comparable with, or in excess of, those seen in Europeans during allergic or parasitic immune responses. Little is known of atopic disease in the Aboriginal population and it has been assumed that elevated IgE values represent anti-parasite reponses. It is possible, though, that an allergic component may be partly responsible and this might account for some of the respiratory difficulties experienced by many Aboriginal people. This is clearly a problem requiring further investigation. No information is available concerning age or gender influences on IgE levels; our data do not suggest any obvious relationships. Perhaps the magnitude of the responses is such that, having exceeded some threshold value, the absolute level might no longer reflect the extent of the disease, hence the clustering of values in the 1000–3000 I.U. ml^{-1} range. Since almost all values were high relative to European means, this could account for the failure of the analyses to demonstrate clear relationships.

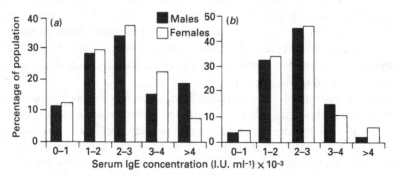

Fig. 14.4 Frequency of individuals with varying concentrations of serum IgE from (*a*) Yirrkala settlement and (*b*) Yirrkala homeland centres.

Immunological parameters: specific antibody responses

It is well established that most, perhaps all, parasitic infections elevate Ig levels: for instance, all classes of antibody are produced against helminths, with IgE of particular significance (Lloyd & Soulsby, 1988). Helminthiases also provide an index of the level of hygiene and sanitation in the community since they depend for their dispersal on the indiscriminate deposition of faecal material on the ground. Do the locality-specific patterns of Ig concentrations found in this study reflect differences in parasite infestation, which in turn relate to differences in residential density and community size and hence in hygiene and sanitation? For each of IgM, IgG and IgA, outstation levels are on average higher than for the settlement, taking into account age and sex effects. It might reasonably be argued that the differences in sewage disposal and quality of water facilities explain the observations. In earlier studies at Donydji homeland centre, over 50% of the residents examined showed evidence of eosinophilia, common in helminthic infection, and the examination of single stool samples by Professor J. Forsyth (University of Melbourne) and Mr S. Pearson (Repatriation General Hospital, Heidelberg) revealed a number of helminthic parasites (as well as cestode and protozoan parasites) (see Table 14.3) in many residents. As most such parasites are difficult to detect in single stool samples, the results are likely to be gross underestimates of real frequencies. With this in mind, sera collected in the present study were tested for antibodies to *Toxocara canis* and *Strongyloides stercoralis* using enzyme-linked immunosorbent assays (ELISA) developed in the Western Australian State Health Laboratories by Mr Neill Hodges and Mr Ian Sampson. The availability of these tests provided an opportunity to assess the distribution of two of the commonly detected helminths previously observed in the Donydji study. The logistic regression analysis, however, did not detect a significant locality effect for these two organisms when comparing township and bush camps, although a higher frequency of *Strongyloides* antibodies was found in the inland camps relative to the coastal locations. We believe that this difference relates to the more efficient disposal of faecal material in the sandy soil of the coastal community and to reduced survival of the parasite.

The transmission of *Toxocara canis* larvae from the faeces of puppies or young dogs to children might not be expected to show either a locality effect or a sex effect, since dogs are ubiquitous in the communities and all children play with them, although there is likely to be greater incidence of cross-infection among schoolchildren in the main settlement. The observed decrease in antibody frequency with age (for males) and the lower frequency of infected males compared with females could be explained in terms of both greater exposure of women to camp dogs (men spending more time away from camp) and perhaps the nutritional disadvantage mentioned earlier. Other Toxocaridae (*T. cati, T. vitulorum* and *T. leonima*, transmitted by cats, cattle and felids or canids, respectively) may also be important in some communities, but cannot be serologically identified at present; their role in the measured responses is unclear. Similarly, the *Toxocara* responses in general do not of themselves explain the relationships of any of the Igs described above, especially as the duration of antibody persistence is unknown (Nicholas *et al.*, 1986).

Strongyloidiasis is endemic to Aborigines in northern Australia, although the incidence reported here (Table 14.3) is significantly greater than the 6% reported previously (W. A. Public Health and Enteric Diseases Unit, 1984). No statistically significant findings emerged from analysis of antibody responses to *Strongyloides stercoralis* except that the incidence of infection was higher in inland bush camps than in coastal camps, possibly for reasons described above relating to the reservoir of parasites in soil. A trend towards lower antibody levels with age raises the question of antibody lability and the relationship of antibody to infection. It is unclear whether antibody titre might be a measure of infection by generation (that is, the young might be more likely to be infected now than 20 years ago) or whether infection rate is constant and antibody titre or response decreases with age. Certainly antibody titre is not an indication of present infection nor of worm burden (Sampson & Groves, 1987). Again, the contribution of antibodies to *Strongyloides* to the overall immunoglobulin patterns is uncertain.

At the request of the community councils, sera were assessed for antibodies to hepatitis B surface antigen (anti-HBs) and for the presence of the antigen itself (HBsAg). The total number of individuals who had serological evidence of infection (immune and chronic carriers) was 338 out of a sample of 403 (83.8%). This is higher than the figure reported recently for a Western Australian Aboriginal population (Holman *et al.*, 1987) and is comparable with the highest reported frequency anywhere in the world. Perhaps not surprisingly, then, no significant locality, age or sex effects were observed, although among chronic carriers more male carriers were found in the settlement but the majority of female carriers and antibody-positive women were outstation residents. Holman *et al.* (1987) found a higher incidence of hepatitis B infection in isolated communities and an increase with age, neither of which could be demonstrated here. The very high frequency of antibodies to hepatitis B is very likely to contribute to the elevated Igs reported earlier, but the distribution does not clarify the relationships observed. The frequency of hepatitis B might also account for the high secondary sex ratio in the Yolngu (White *et al.*, 1990) since there is

Table 14.3. *Parasitic and other infectious organisms of medical importance identified in the population*

Species	Common Name	Identification	Frequency (%)
Helminths (nematodes)			
Strongyloides stercoralis	roundworm	stools	41[a]
		serum antibody	59.6
Toxocara canis	roundworm	serum antibody	43.2
Ankylostoma duodenale	hookworm	stools	45[a]
Trichuris trichiura	whipworm	stools	48[a]
Helminth (cestode)			
Hymenolepis nana	tapeworm	stools	
Protozoa			
Giardia lamblia		stools	
Chilomastix mesnili		stools	
Entamoeba coli		stools	
Entamoeba histolytica		stools	
Endolimax nana		stools	
Iodamoeba butschlii		stools	
Bacteria			
Mycobacterium leprae		medical examination	
Treponema pallidum		medical examination/ serum antibody	
Campylobacter pylori		serum antibody	0.7
Other			
Sarcoptes scabiei var. *hominis*		medical examination	
Pediculus humanus capitis		medical examination	
Dermatomycoses e.g. *Tinea* spp.		medical examination	
Hepatitis B virus		serum antibody	83.3

[a] Identification from single stool samples from 29 individuals at Donydji outstation. Serum antibody determinations from Yirrkala and outstations (see text).

evidence that female carriers of hepatitis B have a higher offspring sex ratio than non-carriers and uninfected mothers (Drew *et al.*, 1986). Of interest too is the observation made by Dr Louise Irving of the Infectious Diseases Hospital, Fairfield, who carried out these assays, that there was a total absence of infection with the defective (hepatitis B virus-dependent) delta virus which, on superinfecting HBsAg carriers, produces progressive and fulminating hepatitis. While delta virus is relatively common in, for instance, Italy, it is not common in southeast Asia; the results suggest that the source of Aboriginal hepatitis might not have been European.

Also at the request of the Aboriginal Council of Yirrkala, sera were screened for anti-treponemal antibodies (TPHA) in order to assess the incidence of syphilis. No attempt was made to distinguish stages of the disease, nor were re-infections identified in the analysis. The results indicate an overall incidence similar to that reported in other Aboriginal communities and they will not be presented in detail here. It can be said, however, that a strong locality effect was evident: a significantly higher frequency of TPHA-positive individuals was found in the homeland centres, a surprising finding in view of the more favourable environment for 'clandestine sexual encounters' provided by the township (Jacobs, 1981). In keeping with the anticipated sexual behaviour, however, there is a higher frequency in older individuals, while among young people more males were infected than females and more infected individuals were found in the settlement. Overall the data are consistent with the possibility that TPHA frequencies might contribute to the generally higher outstation Ig levels described earlier and to the higher levels of Ig found in older individuals.

During the course of TPHA screening by Dr B. Dwyer of the Infectious Diseases Hospital, Fairfield, some sera were screened for IgG antibodies to *Campylobacter pylori*, the organism thought by some to be associated with the aetiology of duodenal ulcers. Only 2 out of 274 (0.7%) Aboriginal people had detectable antibody levels; both were over 50 years of age (Dwyer *et al.*, 1988). The results are consistent with the infrequent observation of peptic ulcers in Aborigines, but raise questions as to its possible emergence in the future, especially if the low antibody frequency represents only recent exposure to the organism.

Conclusions

Any extrapolation from modern traditionally oriented Aboriginal people to those living here prior to European contact must be made with some caution. It would appear, however, that the consequences of Westernisation can be seen in the differences between settlement dwellers and those people resident for longer periods in homeland centres or bush camps, while these differences are blurred by the movement of people between camps and between settlement and camps. The studies centred in the Donydji outstation, which retains its essentially traditional nature, support this view. Many biomedical parameters distinguish settlement from bush residents, in particular the humoral immune response. Elevated immuno-globulin levels in less urbanised individuals (taking into account age and sex effects) probably reflect a greater exposure to infectious agents, many of

Table 14.4. *Summary of statistically significant[a] associations between specific antibody responses, locality, age and sex*

	Specific antibody			
Variable	*Toxocara* sp.	*Strongyloides* sp.	HBs	THPA
Locality				
H = homeland centres	No locality effect	H(In) > H(Co)[c]	H < S(F)	H > S(A,F)
S = settlement				H < S(C)
Age[b]				
C = children or young adults	C > A(M)	No age effect	No age effect	C < A
A = older adults				
Sex				
F = female	F > M(A)	No gender difference	F > M(S)	
M = male			F < M(H)	M > F(C)

[a] Associations based on logistic regression analysis, three-way ANOVA and contingency χ^2 analysis.
[b] C = individuals up to 25 years of age; A = individuals over 25 years of age.
[c] H (In) = inland homeland centres; H (Co) = coastal homeland centres.

which are parasitic, although the specific responses so far measured do not explain fully the observable effects. Although it is not possible to screen large numbers of people for responses to all such organisms, stool samples from individuals in the Donydji outstation indicate that as many as seven or more different parasites may be present simultaneously. In spite of these parasite loads, other biomedical indicators suggest relatively good health and little indication of predisposing factors to the 'lifestyle diseases'. Settlement residents, on the other hand, exhibit reduced immune responses, perhaps indicative of reduced infection, but increased risk of diabetes and cardiovascular disease. Westernisation thus provides considerable improvements in health care, sanitation and hygiene and hence reduces the levels of infection and its consequences. The price paid by people adopting (voluntarily or otherwise) the Western lifestyle is, however, considerable in terms of their susceptibility to other illnesses. The provision of better sanitation and water facilities in bush camps and the availability of health care in more distant places would allow, or perhaps encourage, Aboriginal people to resume more traditional living. Education regarding the inherent problems associated with reduced activity, Western diets and use of substances such as alcohol and tobacco could also lessen the burden on those who assume an urban existence.

Acknowledgements

The authors express their gratitude to the Donydji community for their many years of friendship and support, in particular Roger Yilarama and Tom Gunaminy; to the Aboriginal health workers at Yirrkala, especially Mananu Munungurr, Bandiyal Munungurr and Garnarr Yunupingu; to the members of the Dhanbul Council; and to the hundreds of people who participated so cheerfully. In addition to the generous help of all those mentioned in the text, in particular Dr Kel Semmens and Dr Alicia Polakiewicz, we also thank nurse educator Rob Amery, research student Caroline Jones, and research assistants Eileen Oliver, Sue Hisheh, Liz Hayes and Pam Bagnall, and we are indebted to Louise Bernardi, Helen Skene and Maria Tarzia for their patience and endurance in typing the manuscript. The Australian Institute of Aboriginal and Torres Strait Islander Studies most generously funded much of this work.

References

Drew, T. S., Blumberg, B. S. & Robert-Lamblin, J. (1986). Hepatitis B virus and sex ratio of offspring in East Greenland. *Human Biology* **58** 115–20.

Dwyer, B., Nanxiong, S., Kaldor, J., Tee, W., Lambert, J., Luppino, M. & Flannery, G. R. (1988). Antibody response to *Campylobacter pylori* in an ethnic group lacking peptic ulceration. *Scandinavian Journal of Infectious Disease* **20**, 63–8.

Flannery, G. R. (1985). Changing Aboriginal health profiles in north-east Arnhem Land: Immune status in the transition from hunting and gathering to urbanisation. *Australian Aboriginal Studies* **1**, 52–7.

Gershwin, M. E., Beach, R. S. & Hurley, L. S. (eds) (1985). *Nutrition and Immunity.* London: Academic Press.

Groves, D. I., McGregor, A. & Forbes, I. J. (1975). Impaired immunity in Papua New Guinea highlanders. *Papua New Guinea Medical Journal* **18**, 5–7.

Holman, C. D. J., Bucens, M. R., Quadros, C. F. & Reid, P. M. (1987). Occurrence and distribution of hepatitis B infection in the Aboriginal population of Western Australia. *Australia and New Zealand Journal of Medicine* **17**, 518–25.

Jacobs, D. S. (1981). A syphilis epidemic in a Northern Territory Aboriginal community. *Medical Journal of Australia* (special supplement) **1**, 5–8.

Jones, C. (1990). Fatness and fat patterning among the Yolngu of North East Arnhem Land. In *Is Our Future Limited by Our Past?* (ed. L. Friedman). *Proceedings of the Australasian Society for Human Biology* **3**, 97–112.

Kelley, K. W. (1980). Stress and immune function: a bibliographic review. *Annales de Recherche Veterinaire* **11**, 445–78.

Lloyd, S. & Soulsby, E. J. L. (1988). Immunological responses of the host. In *Parasitology in Focus* (ed. H. Melhorn), pp. 619–30. Berlin: Springer-Verlag.

Nicholas, W. L., Stewart, A. C. & Walker, J. C. (1986). Toxocariasis: a serological survey of blood donors in the Australian Capital Territory together with observations on the risks of infection. *Transactions of the Royal Society of Tropical Medicine and Hygiene* **80**, 217–21.

O'Dea, K. (1984). Marked improvement in carbohydrate and lipid metabolism in diabetic Australian aborigines following temporary reversion to traditional lifestyle. *Diabetes* **33**, 596–603.

O'Dea, K., White, N. G. & Sinclair, A. (1988). An investigation of nutrition-related risk factors in an isolated Aboriginal community in Northern Australia: advantages of a traditionally-oriented lifestyle.' *Medical Journal of Australia* **148**, 177–80.

Rawnsley, H. M., Yonan, V. L. & Reinhold, J. F. (1956). Serum protein concentrations in the North American Negroid. *Science* **123**, 991–2.

Sampson, I. A. & Groves, D. I. (1987). Strongyloidiasis is endemic in another Australian population group: Indochinese immigrants. *Medical Journal of Australia* **146**, 580–2.

Thomson, N. J. (1991). Recent trends in Aboriginal mortality. *Medical Journal of Australia* **154**, 235–9.

W.A. Public Health and Enteric Diseases Unit (1984). *Annual Report*. Perth: State Health Laboratory Service.

Watson, D. S. & Tozer, R. A. (1986). Anaemia in Yirrkala. *Medical Journal of Australia* **144** (special supplement), 513–15.

White, N. G. (1976). A preliminary account of the correspondence among genetic, linguistic, social and topographic divisions in Arnhem Land, Australia. *Mankind* **10**, 240–7.

White, N. G. (1978). A human ecology research project in the Arnhem Land Region: an outline. *Australian Institute of Aboriginal Studies Newsletter* (new series) no. 9, pp. 39–52.

White, N. G. (1985). Sex differences in Australian Aboriginal subsistence; possible implications for the biology of hunter-gatherers. In *Human Sexual Dimorphism* (ed. J. Ghesquiere, R. D. Martin & F. Newcombe), pp. 323–61. London: Taylor & Francis.

White, N. G., Meehan, B., Hiatt, L. & Jones, R. (1990). Demography of contemporary hunter-gatherers: lessons from Arnhem Land. In *Hunter-Gatherer Demography* (ed. B. Meehan & N. G. White), pp. 171–85. University of Sydney.

Wilkinson, G. K., Day, A. J., Peters, J. A. & Casley-Smith, J. R. (1958). Serum proteins of Some Central and South Australian Aborigines. *Medical Journal of Australia* **2**, 158–60.

Zegers, B. J. M., Stoop, J. W., Reerink-Brongers, E. E., Sander, P. C., Aalberse, R. C. & Ballieux, R. E. (1975). Serum immunoglobulins in healthy children and adults. Levels of the five classes expressed in international units per millilitre. *Clinica Chimica Acta* **65**, 319–29.

15 Amerindians and the price of modernisation

K. M. WEISS, A. V. BUCHANAN, R. VALDEZ, J. H. MOORE AND
J. CAMPBELL

Amerindian and genetically related populations in North America currently are affected by a pandemic of obesity, diabetes, and gallbladder disease.
This has arisen in just the past 40–50 years; the conditions were previously
rare in Amerindian populations. The rapid development of the pandemic
thus implicates changes in lifestyle, very likely involving dietary patterns
such as excess calories or newly adopted foodstuffs, which have occurred
on a continental scale. 'Urbanisation', even in small settlements, seems to
be associated with dramatically increased risk. The evidence also suggests
that the conditions have a genetic basis; that is, there is an interaction
between genetic susceptibility and the environmental risk factors. In 1984
Weiss, Ferrell, and Hanis synthesised a variety of indirect pieces of
evidence to hypothesise that these conditions constitute a single genetic
entity, or 'syndrome', that is, are biologically interconnected conditions
sharing genetic risk factors. Subsequent research reports have been
consistent with that idea. In this paper we review some of that evidence, and
present new supportive results from studies in Mexican Mayan and
Mvskoke Creek Amerindians. However, no gene or genes that are
responsible and would prove the syndrome hypothesis have yet been
identified.

Introduction

After World War II, numerous reports began to appear documenting a
rapidly increasing prevalence of non-insulin-dependent diabetes mellitus
(NIDDM) in many groups of Amerindians. Tribal groups from essentially
every part of North America were affected, and there were suggestions that
the same problems existed in Central and South America. Most authors
agreed that this epidemic was associated with a concurrent rise in obesity in
Amerindian peoples.

At the same time, on a similar geographic scale, a rise in gallbladder
disease (GBD) due to cholesterol gallstones was reported. Other reports

221

noted a rise in gallbladder cancer (GBCA), for which the major risk factor is gallstones. Although a variety of risk factors is known to be associated with GBD in Amerindians as well as in other populations, no specific environmental change has been identified that can explain this epidemic in Amerindians.

These conditions now constitute a chronic pandemic in Amerindians; over time, prevalence has been increasing and age of onset decreasing. Several authors have noted that the extremely high prevalence of these diseases in Amerindians suggests that genetic factors are involved (see, for example, Weiss *et al.*, 1984*a*), but the prevailing view has been that the causes are basically environmental and that the conditions are unrelated except insofar as obesity is associated with – and probably a risk factor for – both NIDDM and GBD.

The literature on the epidemiology of these diseases in Amerindian and related populations was summarized by Weiss *et al.* (1984*b*). We argued then that this pandemic may constitute an instance of the effect of 'modernisation' of lifestyles on a gene pool unable to react without pathological consequences. The specific risk factors associated with modernisation have yet to be identified, but we use the term to represent our intuitive sense of a people's participation in the dietary, physical activity, and other aspects of industrial society; there may be specific risks of doing this at a lower socioeconomic status, such as disproportionate use of low-cost but high-fat or high-calorie foodstuffs, or particularly inactive lifestyles.

Specifically, we hypothesised that this is a problem of cultural transition in which (1) these conditions are aetiologically and physiologically related and therefore constitute a complex, or 'syndrome'; (2) the syndrome is due to genotype–environment interaction; (3) the susceptibility allele(s) were in very high frequency in most Amerindian groups (the details of the distribution may reflect the settlement history of the New World); and (4) the closest contemporary Asian relatives of Amerindians do not seem to share the susceptibility, suggesting that it may have evolved early in the settlement of the New World.

We speculated as to the nature of the physiological processing underlying this hypothesized New World syndrome (NWS). In particular, it seemed that lipid physiology might be involved because of the known associations between obesity and both NIDDM and GBD, and because cholesterol gallstones may involve hepatic lipid physiology. These diseases also share clinical and epidemiological associations in other populations and are thought to share jointly a relationship to lipid physiology and energy storage.

In this paper we briefly describe the epidemiology of the components of the syndrome in Amerindians in general, and present new data from two exploratory studies of the New World syndrome. We suggest, more formally than has been done before, a genetic model that may help to formulate analytic strategies with which to attack this problem.

New data

The Oklahoma Mvskoke Creek study

In 1988 we initiated a study of the epidemiology and genetics of these conditions in Mvskoke Indians in eastern Oklahoma. We randomly selected 50 women and their first-degree relatives and spouses, half from rural and half from more urban (small town) subdivisions of the population. A total of 268 individuals were sampled: 25 rural and 25 urban probands, 15 spouses and 203 first-degree relatives. Individuals were interviewed for health and behavioural characteristics, including family history, diabetes and gallstone history, diet, and current medication status. Anthropometric measures were made. A 20 ml whole blood sample was obtained, from which plasma glucose and lipid levels were assayed, and from which DNA for genotyping was isolated.

As shown in Fig. 15.1, the prevalence of NIDDM rises with age in the Mvskoke, and is highest in women over 45 and men over 55. Prevalence is very high compared with the general US population. Table 15.1 compares

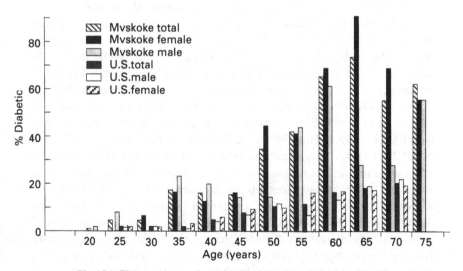

Fig. 15.1 The prevalence of non-insulin-dependent diabetes mellitus by age and sex for U.S. and Mvskoke populations.

Table 15.1. *Prevalence of NIDDM in the Mvskoke, Pima, and Mexican Americans in Starr County, Texas, by age and sex*

	Mvskoke		Pima		Starr County	
Age	male	female	male	female	male	female
15–25	0	0	2.3	1.9	0	0
25–35	4.0	8.0	20.2	14.5	1.2	4.3
35–45	16.7	9.4	34.4	35.1	4.8	7.2
45–55	13.3	33.3	40.2	48.4	9.5	13.8
55–65	53.3	45.8	39.6	68.3	15.6	16.8
65–75	33.3	72.7	40.5	47.5	19.4	16.7
75–85	50.0	40.0	34.2	53.9	—	—

NIDDM prevalence in Mexican-Americans of Starr County (Hanis *et al.*, 1983) with the Pima of Arizona (see, for example, Knowler *et al.*, 1990) and with the Mvskoke. In all these groups the prevalence of NIDDM rises with age, but more so in the Amerindian populations than in Mexican-American populations, in which prevalence is consistent with the level of Amerindian admixture.

Prevalence of each of the components of the NWS and any one or more of the components is shown in Table 15.2. For this study, we define obesity as a body mass index (wt(kg)/ht(m)^2) of 27 or greater; although usage varies, this is a standard cutoff value. At all ages, the majority of the Mvskoke study population is overweight, and prevalence of obesity does not rise with age. Even in the youngest age group half are already overweight. Although obesity is often associated with NIDDM, or with increased severity of disease, this is not always so. In the Mvskoke, prevalence of NIDDM is higher in women whose body mass index (BMI) is at or over 27; 56% of women over age 45 have NIDDM compared with 30% of women with BMI less than 27. But among men over age 45, 36% with BMI less than 27 and 36% with BMI greater than or equal to 27 have NIDDM. Obesity is not necessarily a precursor of NIDDM, especially when measured without regard for its distribution within the body, but in Mvskoke women obesity seems to be associated with increased risk.

Current evidence shows that abdominal obesity is a risk factor for diabetes. To test this association in the Mvskoke, a standardized index of abdominal obesity was developed (Valdez, 1991). This index, the Conicity Index, is the contrast between the waist of a given person and the 'waist' of an imaginary cylinder of the same height and weight. The more the waist of that person departs from that of the cylinder, the closer his or her waist will

Table 15.2. *Prevalence of single conditions, and of syndrome, by age and sex*

Age	Diabetic %	N[b]	BMI ≥ 27 %	N	Gallbladder Disease %	N	Syndrome[a] %	N
Male								
15–25	0	6	50.0	2	0	6	50.0	2
25–35	4.0	25	50.0	12	0	26	63.6	11
35–45	16.7	18	50.0	10	5.6	18	70.0	10
45–55	13.3	15	63.6	11	0	15	70.0	9
55–65	53.3	15	50.0	12	7.1	14	91.7	12
65–75	33.3	12	44.0	9	41.7	12	88.9	9
75–85	50.0	2	0	1	50.0	2	100.0	1
Female								
15–25	0	19	72.7	11	5.3	19	72.7	11
25–35	8.0	25	50.0	20	8.3	24	55.0	20
35–45	9.4	32	64.0	25	25.0	32	80.0	25
45–55	33.3	33	71.4	28	27.3	33	89.3	28
55–65	45.8	24	54.6	22	33.3	24	86.4	22
65–75	72.7	22	73.7	19	33.3	21	100.0	19
75–85	40.0	10	60.0	10	40.0	10	90.0	10

[a] The three preceding columns do not necessarily sum to the syndrome total, as individuals may have more than one of the conditions.
[b] Total number of subjects for whom the value of this variable is known.

be to the waist expected for a double cone. The values of this conicity index ranged from 1.0 to 1.5 in this population. The median in this sample was 1.26, and 43% of the individuals with values higher than the median were diabetic, but only 15% of the individuals with conicity index lower than or equal to the median were diabetic (R. Valdez, unpublished data).

GBD prevalence is probably very conservatively estimated because it is based on self-reported disease. The definitive test, an ultrasonographic scan, was not possible under the field conditions of this study, so cases of undiagnosed, silent gallstones are not included.

The final column in Table 15.2 shows the prevalence of any or all of the three conditions of interest; NIDDM, GBD or body mass index ≥ 27. An individual is considered affected if he or she has one or more of these conditions. At least half of each age group, both men and women, is affected, and in the older ages, the proportion affected approaches or reaches 100%.

We found no significant difference in disease rates by place of residence (rural or urban) when age was controlled. In the rural segment, 51.3% of

15–45 year olds were affected with at least one of these conditions, and 72.5% of those over 45 were affected. In the urban, more acculturated segment, 38.7% of the younger group and 77.1% of the older group were affected. Since admixture and urbanisation are both known risk factors, we believe that there are two possible explanations for these results: (1) the urban–rural distinction does not represent enough genetic or lifestyle variation to show an effect on disease risk; or (2) admixture and lifestyle do vary between the urban and rural groups but are countervailing risk factors so that the net effect on disease prevalence is similar.

Polymorphisms for several potentially relevant genes ('candidate gene loci') were tested in a subsample of unrelated Mvskoke individuals, but although effects on lipids and other measures were identified, no effect close to the strength required to explain these disease problems in the Mvskoke was found (R. Valdez, in preparation). The loci included the insulin receptor, two glucose transporter genes, apolipoprotein B and the apoAI/CII/AIV gene cluster, and the gene for the enzyme lipoprotein lipase. Similarly, plasma lipids were assayed, but no patterns of major explanatory value relative to the general problem of NIDDM in Amerindians emerged; the Mvskoke showed elevated triglycerides and somewhat elevated LDL when diabetic, but otherwise their lipid levels were low compared with Caucasians with the same level of obesity (P. Kris-Etherton *et al.*, unpublished data).

Mexican Mayans

Between 1984 and 1987, pilot survey studies were done among the Mayan populations of the Yucatan peninsula of Mexico. Small numbers of Mayans were sampled from several groups representing a range of acculturation and were measured for a variety of relevant traits. The groups included recent immigrants to a low-income peri-urban neighbourhood of a city of 150 000, several villages of 500–5000 within 50 km of the city, and two villages in the remote part of the Yucatan jungle, many miles from easy contact with any urban area.

The data collected included basic anthropometric measures, a short food-frequency questionnaire, and a 10–20 ml sample of whole blood from most subjects. Basic glucose and plasma lipids were assayed; results are shown in Table 15.3. It is clear that even modest modernisation has serious metabolic effects. In the remote villages, which have relatively infrequent direct contact with the outside economy, obesity, elevated glucose, diabetes, and GBD were rare. Generally, in the villages with more contact with urban areas, all the NWS indicators (weight and blood glucose, for example) were elevated. Dietary and lifestyle questions, however, failed to

Table 15.3. *Mean phenotypes from sampled lowland Mexican Mayan populations*

(*a*) Chronic disease risk factors

	Age	B.M.I.	Height (cm)	Weight (kg)	Chol	Gluc	Hb$_{A1C}$	N	Pop
Isolated villages									
1	35.2	23.7	144.2	51.08	142.9	100.6	*	58	500
2	31.8	24.0	143.6	49.67	160.4	—	4.1	15	400
Slightly acculturated									
3	33.8	27.8	145.2	58.15	174.8	117.3	4.8	17	299
More urbanised									
4	40.6	27.3	145.4	57.29	169.3	93.6	5.6	39	3486
5	31.9	26.6	146.2	56.70	127.9	148.5	*	33	5000 +

Values are sample means; BMI is the body mass index = weight (kg)/height (m)^2; Chol is the serum concentration of cholesterol (mg per 100 ml); Gluc is the concentration of glucose in venous serum from a 4 + hour fasting sample (mg per 100 ml); Hb$_{A1C}$ is percent glycosylated haemoglobin (Glyc-affin test).
* Assay not yet completed.

(*b*) Prevalence of diabetes or suspected diabetes in Mayan populations and in comparison data

Population	Confirmed diabetes		Suspected diabetes		Total
	no.	%	no.	%	(%)
Isolated populations					
1	0	0	0	0	0
2	0	0	1	01.6	01.6
Slightly acculturated					
6	3	12.5	1	4.2	16.7
7	4	22.3	2	11.2	33.5
3	2	13.3	2	13.3	26.6
More urbanised and acculturated					
4	7	17.9	8	20.5	38.4
5	10	30.3	10	30.3	60.6
U.S. high-risk populations					
Starr Co, Texas	—	6.9	—	—	—
Laredo, Texas	—	10.5	—	—	—
San Antonio, Texas	—	9.5	—	—	—
Pima Indians	—	24.3	—	—	—
U.S. Caucasians	—	2.5	—	—	—

Confirmed diabetes is by physician's diagnosis and/or biological test (serum glucose > 120; HB $_{A1c}$ > 8%). Suspected diabetes is HB A$_{1c}$ > 7.5%, glucose > 100, or multiple diabetes symptoms. Village names are not used to protect anonymity; two additional villages are reported in (*b*) that are not included in (*a*).
Source: E. Georges & K. M. Weiss, unpublished data.

elicit any indication of what the responsible factors might be. However, even these small samples and cursory data show the high sensitivity of the Mayans to the diabetogenic effects of even a relatively low level of 'modernisation'.

Current status of the New World 'syndrome' hypothesis

The following section is a brief digest of recent literature concerning traits that Weiss *et al.* (1984*a*) suggested are components of the New World 'syndrome', as that literature relates to the current level of support for the syndrome hypothesis. The points stressed are the specific susceptibility of Amerindian and related populations to these problems, and the associations among the problems themselves that suggest they may have common underlying risk factors. These points are elaborated in Weiss *et al.* (1993).

Diabetes: genetic epidemiology

A substantial number of recent papers have continued to document the high level and continental distribution of NIDDM in Amerindian populations, and in admixed Hispanics such as Mexican-Americans, consistent with a genetic aetiology. Although no specific genes have yet been identified, there is considerable evidence for genetic susceptibility within the high-risk populations. In Mexican-Americans (Hanis *et al.*, 1983) and in the Pima (Knowler *et al.*, 1988), parental disease state is correlated positively with risk in offspring. In the Pima, an individual with two affected parents is at higher risk than an individual with one affected parent. Similar results have been found in a survey of Oklahoma Indians (Lee *et al.*, 1985) and Canadian Indians in Southwestern Ontario (Evers *et al.*, 1987).

Recent reports from the Pima show that the cumulative incidence of diabetes is greater and increases more rapidly as a function of parental diabetes status and age of onset (Knowler *et al.*, 1988). Severity of parental hyperglycaemia had similar effects. The same data show a positive association between obesity, severe parental diabetes and early age of onset, and the age-adjusted incidence of NIDDM.

The search for specific genetic markers has found several leads but none with the strength of association to be the major locus in question. In the Pima, associations with alleles of the HLA-A genes, as well as general evidence consistent with major locus effects, have been found; the general evidence from Amerindians is reviewed by Szathmary (1987). In Mexican-Americans there have been reports of effects at the Rh system and the haptoglobin locus, but these have not been consistent. A study of seven gene loci in material from Mexican-Americans in Colorado, including haptoglobin, has found no association (Iyengar *et al.*, 1989). As noted earlier, no substantial effects were found for five candidate loci in the Mvskoke.

A number of specific genes or physiological traits that are directly related to diabetes physiology may be variable in Amerindian or related populations. Insulin action has a trimodal distribution in Mexican-Americans and is heritable (Lillioja *et al.*, 1987), and there is an association between NIDDM and an RFLP at the insulin receptor gene that may be Amerindian-specific (Raboudi *et al.*, 1989). Studies of insulin response to an oral glucose load show a similar pattern in the Navajo and in the relatively unacculturated (and diabetes-free) Dogrib of northern Canada – two groups previously thought to be relatively diabetes-resistant compared with other Amerindians – but quite different from the response pattern in Caucasians (see, for example, Szathmary, 1989). However, these variants do not appear to be sufficient to account for the existence of elevated susceptibility in Amerindians. Rather, they may be the kind of modifying genes one would expect within any population.

Accumulating admixture data continue to support the argument that these problems are at least partially genetic in nature (Chakraborty, 1986; Chakraborty & Weiss, 1986). In the Pima, a European genetic marker at the Gm system of immunoglobin is associated negatively with prevalence and positively with onset age of NIDDM (Knowler *et al.*, 1988). There is also a correlation between prevalence and fraction of self-reported 'Indian heritage' in eighths of known ancestry. In the Pima community, full heritage is associated with about double the risk of those with no reported Amerindian ancestry. The Mvskoke also show an association between risk and self-reported fraction of Amerindian ancestry.

In Mexican-Americans, admixture is also associated with prevalence of NIDDM in the direction expected. In their Starr County, Texas, sample of Mexican-Americans, Hanis *et al.* (1990) have estimated 31% Amerindian ancestry, while 61 and 8% of the genes are of Spanish and African origin, respectively. Amerindian admixture roughly corresponds to the lifetime prevalence of NIDDM in that population. Hanis *et al.* (1991) have also estimated an Amerindian admixture of 18% for both Cubans and Puerto Ricans, which is consistent with the level of NIDDM in those populations. A study of non-Hispanic whites and Mexican-Americans in Colorado has also found a higher prevalence of NIDDM in the latter (Hamman *et al.*, 1989).

Obesity

A very high prevalence of obesity is found in many populations that also have high prevalence of NIDDM, and increasing trends in both have occurred in diverse Amerindian and admixed groups in the last 50 years (Knowler *et al.*, 1990; Diehl & Stern, 1989). Associations between obesity and NIDDM are found in the Pima (see Knowler *et al.*, 1990) and Mexican-Americans (Diehl & Stern, 1989; Samet *et al.*, 1988; Mueller *et al.*, 1984; Lee *et al.*, 1985). However, obesity alone does not account for all of the excess risk of NIDDM in Amerindians or Mexican-Americans (see, for example, Haffner *et al.*, 1986*a,b,c*; Haffner, 1987; Knowler *et al.*, 1990).

Much of the investigation of the role of obesity in NIDDM has focused on the association of fat patterns with disease risk factors. Attention has been paid to the ratio of central or truncal adiposity to peripheral adiposity (commonly measured as the subscapular to triceps skinfold ratio and termed the centrality ratio or index, or as the waist to hip circumference ratio). Physiological reasons for this have been suggested (see, for example, Stern & Haffner, 1986).

In the Mvskoke, the frequency of upper body obesity is high. When compared to the distribution of centrality ratios in Mexican-Americans and Caucasians reported by Haffner *et al.* (1989), the fraction of Mvskoke men in the upper tertile of the centrality distribution described by Haffner *et al.* (1989) (> 1.92) is much higher than in Mexican-Americans or Caucasians (70% compared with 35 and 27% repectively). The proportion of Mvskoke women in the top tertile (> 1.21) is also much higher; 64% compared with 30 and 21%, respectively. Fifteen per cent of Mvskoke men fall in the low tertile (< 1.5) compared with 31% of Mexican-Americans and 41% of non-Hispanic whites (NHW); 17% of Mvskoke women are in the low tertile (< 0.95) compared with 28% of Mexican-American women and 53% of Caucasian women. Stern and Haffner (1986) report that in Mexican-American women in the San Antonio (Texas) Heart Study, the centrality index was significantly correlated with percent Native American genetic admixture ($p < 0.001$). This was not true for men, however.

In the Mvskoke, the centrality index is not a good predictor of NIDDM. Diabetic men have a mean centrality ratio of 2.9 compared with 2.6 in non-diabetic men; diabetic and non-diabetic women alike have a mean ratio of 1.6. These differences are not statistically significant. This finding suggests to us that if central obesity is a risk factor for NIDDM then many people unaffected in this study will become affected at a later time, as much of the population already has high-risk obesity patterns. On the other hand, as stated earlier, abdominal obesity may be a good indicator of diabetes; 74% of Mvskoke diabetics have a conicity index that is above the median for the sample, whereas only 40% of the non-diabetics are above the median. This 40%, we think, is the group at risk of developing diabetes.

Gallbladder disease

The excess risk of gallbladder disease in Amerindian populations has been clearly documented, and the evidence that GBD in Amerindians is a genetic problem was summarised by Weiss *et al.*, (1984*a,b* and references therein). In the Pima, there is an increasing prevalence of GBD with time, particularly as the self-reported fraction of Amerindian ancestry rises (Knowler *et al.*, 1984). Relative risk was not found to be higher in women with affected relatives, but this is not unexpected given the very high prevalence (and gene frequency, if aetiology is genetic). As summarised by Diehl & Stern (1989), GBD is more common in Mexican-Americans than can be explained by parity or obesity alone. In a population of Mexican-Americans in San Antonio, the prevalence of GBD is correlated with the fraction of Amerindian admixture. In a large study of Mexican-Americans in Starr County, Hanis *et al.* (1983) found that first-degree relatives of individuals with clinically symptomatic GBD had 1.8 times the random risk of developing GBD; that this is not due to environmental factors is shown by a lack of an excess risk in spouses of affected people.

A recent study of hospital records from the government health service in Manitoba, Canada, has found that Amerindians experience an overall relative risk of cholecystectomy of 1.46 compared with non-Amerindians (Cohen *et al.*, 1989). A survey of 704 adult Cree–Ojibwa Indians in the subarctic boreal forest of northern Manitoba and Ontario has also revealed a high prevalence of clinical GBD: 18.5% in all women aged 20–64, and 30% in those aged 50–64 (Young & Roche, 1990). Of a variety of predictor variables tested, triglycerides (TG) were positively and total cholesterol (TC) negatively associated with GBD status.

Gallbladder disease is statistically associated with NIDDM in most populations that have been studied, probably owing to the common association with obesity and/or other shared physiological pathways involving lipids, although one study in non-Amerindians has shown that GBD is probably not simply a consequence of NIDDM but that insulin resistance is a powerful risk factor (Laasko *et al.*, 1990). Hanis *et al.* (1991) found that in males the prevalence of gallbladder disease rose to about 25% by age 45, and to over 40% in females (the corresponding percentage for NIDDM was about 15%, and for obesity nearly 30% in males, and over 40% in females). They found highly significant associations of GBD with NIDDM and/or obesity; the strongest associations were for those with the most severe or the earliest symptoms.

There is an association between obesity and risk of gallbladder disease in the Pima (Howard, 1987; Knowler *et al.*, 1984), as well as in Mexican-Americans in Starr County, Texas (Hanis *et al.*, 1985), New Mexico (Samet *et al.*, 1988) and San Antonio (Diehl & Stern, 1989; Haffner *et al.*, 1989). However, the associations are not simple and threshold effects in these very obese populations may sometimes obscure them. Haffner *et al.* (1989) showed that GBD in San Antonio Mexican-Americans is elevated even after adjusting for central body fat deposition. Both BMI and the centrality ratio were positively and independently associated with GBD in women, while in men BMI but not the ratio was associated with gallbladder disease.

Howard (1986) has recently summarised relationships between obesity, cholelithiasis and lipoprotein metabolism, stressing her work in the Pima, where there is a weak relationship between obesity and gallbladder disease, and a positive correlation of 0.53 between BMI and biliary cholesterol – but not bile acid or biliary phospholipid – production. BMI was also correlated with bile cholesterol saturation level. This appears to be due to an increase in total body cholesterol synthesis with obesity. In other studies in the Pima, obesity is positively correlated with very low-density lipoprotein (VLDL), total TG and reduced high density lipoprotein cholesterol (HDL) levels; the latter correlation is mainly due to reduced HDL_2 subfraction levels. Essentially the same lipid profile is associated with NIDDM in non-Amerindians. A nonlinear relationship between cholesterol and GBD history was found in Starr County (Hanis *et al.*, 1985), perhaps suggesting similar effects, but asymptomatic stones were not a part of that study so this remains unresolved. In the Mvskoke, obesity was not indicative of gallstones. Age, sex, and diabetic status were the best indicators of gallstone presence: older diabetic women showed an increased risk for gallstones.

Dietary considerations

There has been much speculation about the possible dietary cause of this problem, specifically about the comparison between aboriginal and modernised diets. Local explanations for the recency of the epidemic have been supported recently by studies showing that traditional Pima foods cause less rise in plasma glucose and insulin than the caloric equivalent of glucose, at least in part because the traditional foods are more slowly digested or absorbed (Brand *et al.*, 1990). Many authorities have attributed these disease problems in Amerindians to generally overcaloric nutrition relative to physical activity (see, for example, West, 1978*a,b*). Weiss *et al.* (1984*a*) speculated that, because the NWS conditions appear to involve lipid physiology, one factor may be some continent-wide change in specific dietary constituents (e.g. vegetable oil substituted for animal fat). Perhaps the early reliance of aboriginal Amerindians on sea food and arctic land mammals, which provide sources of less saturated fats than current animal products, was involved. However, Szathmary (1990) has noted the aboriginal arctic diet was heavily dependent on protein and fat, and poor in sources of carbohydrate; this kind of diet does not provide large sources of quickly accessible glucose, and it may be that genetic variants leading to a more efficient gluconeogenesis may have enjoyed a selective advantage.

It may be stated with confidence that the genetic basis of Amerindian susceptibility to dietary change cannot be attributed solely to adaptation to any *local* Amerindian environment, such as the desert environment of the Pima (Knowler *et al.*, 1983; Pettitt & Knowler, 1988). Such environments may be proximally relevant to the expression of disease, but this is a continental problem and its origins require a continental explanation.

A more formal specification of the genetic model for the NWS

The New World syndrome hypothesis was based, and still rests, on circumstantial evidence and 'geographic epidemiology'. Weiss *et al.* (1984*a*) did not provide a formal statistical model for the hypothesis. The following provides an attempt to specify such a model.

The essence of the model is that: (1) a set of traits, including a tendency to develop obesity, cholesterol-saturated bile, and NIDDM, as well as their pathophysiological sequelae, exists in Amerindian and genetically related populations; (2) there is a common, not too complex, genetic basis for this susceptibility; (3) the development of the traits to a pathological level depends on interaction with some aspects of the environment that have changed in the past 40–50 years; (4) the gene(s) involved have very high allele frequency, which was reached in Amerindians early in the settlement of the New World so that most of their descendants, in diverse tribes, share these alleles identical by descent from Amerindian ancestors; and (5) in addition to the 'major' causal genes, a 'normal' level of variation at ancillary loci ('polygenes') modifies risk among individuals with the same major-locus genotype.

At present, no primary enzymatic or genetic defect responsible for these problems is known, so that standard segregation and linkage analysis cannot be used directly to find the gene(s) responsible. Instead, individuals are classified according to several end-point outcome phenotypes (NIDDM, GBD, obesity) and measured for several quantitative risk-factor phenotypes potentially related to these end-points (e.g. cholesterol and glucose levels). All of these phenotypes are likely only to be indirect manifestations of the primary underlying defect; indeed, the 'syndrome' constitutes the pleiotropic effects of the underlying genotype. Underlying genetic factors must be inferred from the data, and the model is that the outcome phenotype is the result of the major genotypes, plus effects of some ancillary genes and environmental factors.

A powerful standard model for specifying univariate models of this kind is widely available and is known as the 'mixed model' of segregation analysis (Elston, 1980, 1981, 1986). If the outcome phenotype is ϕ, we specify the effects of a major genotype (MG), ancillary genes, or 'polygenes' (PG), unmeasured environmental factors randomly affecting individuals (E), a series of n risk factors not related to the genes in question (x_i), and a series of r risk factors that may have major-genotype-specific effects (y_j). The latter two factors are assumed to have been identified and *measured* or observed on each individual.

Using these terms, the phenotype is determined in the following way (omitting higher-order effects and other complications):

$$\phi = \mu_{MG} + PG + E + \sum_{i=1}^{n} \beta_i x_i + \Sigma \gamma_{i,MG} y_j, \qquad [1]$$

where μ_{MG} is the mean effect on the phenotype of the MGth major genotype. The linear regression term $\Sigma \beta x_i$ represents an adjustment for the set of genotype-non-specific measured risk factors, and the last term is a genotype-specific regression for the effect of the measured risk factors given the MGth major genotype. As an approximation it is usually assumed that the unmeasured polygenic and environmental effects can be parameterized as normally distributed deviations from mean effects of 0, variance σ^2_{PG+E}.

In a population there will be one group for each genotype at the major locus (or loci). In the population, these will be in proportion to their genotype frequency (for example for a two-allele locus with allele frequency p, the groups will have frequency p^2, $2p(1-p)$, and $(1-p)^2$), with group phenotype means μ_{11}, μ_{12}, and μ_{22}, respectively. Within each group, the variance of the phenotype distribution due to unmeasured factors is σ^2_{PG+E}, equivalent to the assumption that all major genotypes respond to these factors in the same way, so there is a burden on us to identify the genotype-specific interacting environmental factors. For any of these means that differ significantly, there will be a separate mode in the phenotype distribution. For example, if the major locus is dominant, $\mu_{11} = \mu_{12}$, and there will be only two groups.

A model for the NWS is an extension of equation [1] to include multivariate outcome phenotypes, either those that we identify as 'diseases', such as obesity, NIDDM, or GBD, or physiological measures we assume to be consequences of the underlying major-gene effects. Dichotomous phenotypes can be modelled by standard modifications of

[1], for example, by logistic models (Bonney, 1986, 1988; Bonney *et al.*, 1988). For each individual these values would be adjusted for age, sex, parity, use of medication, and so on as the x variates in [1]. In addition, relevant environmental exposures such as dietary constituents or physical activity measures should be included as the genotype-specific (y variates).

Equation [1] can be rewritten in matrix form as follows:

$$\phi = \mu_{MG} + PG + E + \beta'x + \gamma'y.$$ [2]

(The notation β' and γ' indicates the transpose of the matrix.) For example, if there are two phenotypes, two measured genotype-non-specific and two measured genotype-specific risk factors (these numbers need not be equal), equation [2] will be

$$(\phi_1,\phi_2) = (\mu_{MG,1},\mu_{MG,2}) + (PG_1,PG_2) + (E_1,E_2)$$

$$+ \begin{bmatrix} \beta_{11} & \beta_{21} \\ \beta_{12} & \beta_{22} \end{bmatrix} \begin{bmatrix} x_1 \\ x_2 \end{bmatrix} + \begin{bmatrix} \gamma_{11} & \gamma_{21} \\ \gamma_{12} & \gamma_{22} \end{bmatrix}_{MG} \begin{bmatrix} y_1 \\ y_2 \end{bmatrix}$$

$$= (\mu_{MG,1},\mu_{MG,2}) + (PG_1,PG_2) + (E_1,E_2)$$

$$+ (\beta_{11}x_1 + \beta_{21}x_2,\beta_{12}x_1 + \beta_{22}x_2) + (\gamma_{11}y_1 + \gamma_{21}y_2,\gamma_{12}y_1 + \gamma_{22}y_2)_{MG},$$ [3]

where the *MG* subscript indicates that the regression coefficients are conditional on (that is, vary for each) major genotype. In the population, conditional on the major genotype, and measured in risk factors,

$$\mathrm{Var}(\phi) = G + E,$$

where each term is a matrix. For Var (ϕ), the diagonals are σ_i^2, the variances of the individual traits, and the off-diagonals are σ_{ij}, the covariances between the traits. For G and E the pattern is the same, except that the values are the contributions of these *effects* to the variances and covariances. We typically assume that as an approximation these effects are additive and multivariate-normally distributed.

This model requires that we identify on all individuals the risk factor that has changed during the post-war period (if the NWS hypothesis is approximately correct). To test the model, and estimate its parameters generally, will require family data (molecular 'tricks' might be possible if the hypothesis is correct as will be mentioned briefly below). In family data, the behaviour of the polygenic and unmeasured environmental factors can be traced using standard segregation analytic methods (various options for such methods exist, including relaxation of the assumptions associated with polygenic inheritance). Several papers can be consulted for applications and methods of these types of models (Lalouel, 1983; Lange & Boehnke, 1983; Bonney & Elston, 1985*a,b*; Blangero *et al.*, 1990; Blangero & Konigsberg, 1991; Bonney *et al.*, 1988).

This model is essentially a standard extension of the existing 'mixed' model for quantitative traits to multivariate correlated phenotypes with gene by environment interaction with identified risk factors. The model in [2] shows only linear regression effects. However, we know that the NWS traits were rare or even absent in Amerindian populations before World War II, approximately. Therefore, associated with the distribution of environmental exposures at that time, *no* genotype produced pathogenic phenotypes, whereas in a Westernised environment all genotypes appear to be susceptible (see, for example, Trowell & Burkitt, 1981), but the Amerindian genotype(s) have a penetrance that may approach 85–100%.

The evidence suggests that, as the distribution of environments has changed, the relative increase in risk associated with Amerindian genotype(s) has been much greater than that in Caucasian (or Black or Asian) genotypes. If this is true then : (1) the assumption of equality of variances among genotypes will be seriously violated if the risk factors are not measured and regressed (as y values in equation [2]); (2) if the appropriate risk factors *are* measured, there may be no residual genotype-specific mean effect; and (3) the regressions in [2] should perhaps be made nonlinear (e.g. $\gamma_{j,MG} y_j^r$ or $\alpha_{j,MG} + \gamma_{j,MG} y_i + \gamma'_{j,MG} y^2$, etc.). Among the variables to be included may be age, birth cohort, or time since exposure to the provocative environmental factors (whatever they are).

A comparison of the available prevalence or admixture data suggests compatibility of the NWS (as reflected in cross-sectional NIDDM prevalence) with a recessive or codominant single-locus model (Chakraborty, 1986; Chakraborty & Weiss, 1986; unpublished other data, B. Mitchell & M. Stern, personal communication). The establishment of such a model must await formal analysis.

If we assume that the same major locus effect is absent in Caucasians, but exists at very high frequency in pure-blood Amerindians, the model in [2] suggests why it has been difficult to detect the major genotypes within tribal groups such as the Pima or our Mvskoke study. If $p \approx 1.0$ in Amerindians, then almost all individuals will have at least one copy of the susceptibility allele(s), and most will be homozygous susceptibles. In the model, little if any phenotypic variance will be associated with the major locus. The many familial patterns and candidate-gene effects reported in the Pima can then be explained as comprising elements of the ancillary (e.g. 'polygenic') variance that is expected within major-locus genotypes. This suggests why study of this problems may well be much more effective in substantially admixed populations, such as Mexican-Americans.

A major reason why studying this problem in an admixed population, such as Mexican-Americans, may provide a critical advantage in understanding this problem is that the allele frequency may be high enough (e.g.

0.3 if it corresponds to the amount of Amerindian admixture) that all genotypes will be common and many families will be segregating both high and low susceptibility alleles, variation necessary to characterize the gene.

There are several practical problems with this model. First, even if we could identify all the relevant risk factors and phenotypes, we are far from having practical computer models that can accommodate all these details. And many of the characters we have been discussing (NIDDM, GBD, obesity) are artificially dichotomised (e.g. affected or unaffected) when in fact they are the result of continuous underlying risk phenotypes; this weakens our inference, and it would be preferable to measure the underlying phenotypes directly. We may not yet have identified all the critical risk factors, but these should not be left in the general environment term in equation [2] because that is not genotype-specific and exposure to such factors is usually assumed to be normally distributed and homoscedastic for all major locus genotypes, which may not be correct.

Thus, while the statistical model presented here may describe the situation properly, special data or analytic 'tricks' may be required to resolve the problem, unless we can identify factors closely connected to the underlying genotypes. The likelihood of [2] can, in principle, be tested in family data (Elston, 1980, 1981, 1986), but will not be very informative if too many family members are uninformative, for example, by being too young. In that case, other methods can be used to apply the same model, for example, to sets of affected relatives.

We referred earlier to molecular methods that may be applicable. On the assumption that the NWS hypothesis is correct, then most Amerindians will share some allele(s) that arose, by drift or positive selection, to very high frequency early in the settlement of the Americas. If so, at the locus concerned, and excepting the accumulation of a few point mutations in and around the locus, Amerindians will be highly homozygous. On the other hand, overall, Amerindians are genetically only slightly less heterozygous than any other racial group. This suggests that a molecular search for regions of shared homozygosity among Amerindians may identify candidate regions for more formal segregation or linkage analysis in families in an admixed population. Exploratory statistical studies suggest that not too many spurious high-homozygosity regions will exist in the 1000 or so generations since the first settlement of the Americas (A. Connor, in preparation). However, that same work also suggests that the homozygous region may be only about 1–5 cM wide, too narrow to detect with current linkage mapping methods. Special molecular searching techniques may be required. We are exploring possibilities along these lines, but no methods are currently available.

Discussion

It is common to hear diseases like obesity, GBD, and NIDDM characterised as 'Western' diseases, because they have become major causes of adult
morbidity and mortality in the industrialised nations (see, for example,
Trowell and Burkitt, 1981). However, it should be noted that many
acculturating peoples suffer these 'Western' diseases to a far greater extent
than do the Western peoples themselves, often while participating only
marginally in 'Western' society. Examples are the Polynesians and
Amerindians and their problems with diabetes (Weiss *et al.*, 1984*a*; Baker,
1984; Trowell & Burkitt, 1981). Although Western environments do clearly
seem to be implicated in these diseases, that is only a part of the story. In
Amerindians, genes appear also to be involved.

The term 'syndrome' is traditionally applied to sets of conditions that are
present in affected individuals and absent otherwise. In the case of the NWS
hypothesis, the conditions (NIDDM, GBD, obesity) are based on rather
arbitrary cutoff criteria for continuous underlying distributions. Thus, in
this case the syndrome reflects the relative probability of being within one
part of a continuous multivariate distribution rather than another, based
on genotype. Reviews of the evidence from the Pima (Howard, 1986, 1987;
Knowler *et al.*, 1990) and Mexican-Americans (Diehl & Stern, 1989) show
that it cannot seriously be disputed that populations with Amerindian
genetic ancestry are suffering a pandemic of NIDDM, GBD, obesity, and
other associated traits, and that these are correlated with each other. In that
sense the NWS is an *epidemiological* fact. It remains to be proven that it is
also a *genetic* fact.

Although some differences have been reported in the pathophysiology of
these traits between Amerindians and other racial groups, the pattern is not
different enough to provide much help in searching for altered physiological pathways in Amerindians. NIDDM, obesity, and GBD appear earlier
and more often in Amerindians, but these traits are epidemiologically
associated in other populations. This suggests that the cause of the
increased susceptibility is simply a different 'timing' of some underlying
enzymatic reaction(s), rather than, for example, a genetically 'deleted'
pathway.

Yet, not all racial groups have the same correlations among these risk
factors (Weiss *et al.*, 1984*a*). For example, at least one susceptible group,
the Polynesians, do not seem to suffer GBD along with their obesity and
NIDDM to the extent that Amerindians do, and in fact GBD is relatively
rare in Black populations (Weiss *et al.*, 1984*a,b*; Lowenfels, 1985; Trowell
& Burkitt, 1981). Also, there are repeated observations that CVD is
markedly reduced in Amerindians and admixed populations, relative to the
level of obesity, NIDDM, and GBD (Diehl & Stern, 1989). This suggests

that the problem is not just *more* of the same in Amerindians but that some aspects of their network of metabolic pathways are used proportionately more (or less) efficiently or readily, in the same environments, relative to other racial groups.

Finding the susceptibility gene requires a formal specification of the statistical model and an explicit attempt at linkage mapping and segregation analysis. We have sketched out the basis of such a model here, and specify more of the model in a forthcoming publication (Weiss *et al.*, 1993). Although mapping of a multivariate quantitative trait is not easy there is nothing that formally precludes this being done successfully. However, the relatively low level of variation in susceptibility within Amerindian populations suggests that segregation of risk – the critical ingredient for linkage mapping – will be easier to find and characterise in admixed populations such as Mexican-Americans. Given the problem of defining the phenotype, in particular the lack of a reliable risk-precursor state, it may be important to use strategies only involving sets of affected relatives, or in some other way to reduce the problem of false-negative phenotype classification.

The candidate genes so far tested are those for which probes have been available and which are known to be involved in physiology related to diabetes. However, if some unidentified underlying gene is involved (or even a locus that regulates the genes that have been tested), these are not good candidates. A genome-screening effort would be warranted; this can now be done by use of hypervariable regions such as dinucleotide repeats and random highly polymorphic DNA marker loci.

Studies of diet and other aspects of lifestyle are important, but until genotypic differences are known these will probably continue to be unproductive. If the changes that are needed to trigger the pathogenic reactions are subtle, it may be that most individuals in Amerindian or admixed populations are exposed. Risk differences associated with dietary differences may mainly reflect the action of other genes, severity differences, or differences in how the underlying genes are expressed, rather than the expression or non-expression of those genes. This is only made worse if there are thresholds above which dietary differences make little difference in that expression. Ironically, the studies that have been done may actually be misleading in this respect. If everyone is exposed to risk factor X that triggers the basic problem, and variation in timing and expression is modified by risk factors Y and Z, the latter may appear to be associated with disease in the population by, for example, being associated with the more severe or earlier cases. Not only would it be wrong to believe that Y and Z were causing the disease, but the prevalence of susceptibility would be underestimated if based on studies of Y and Z. Similar arguments apply to genetic factors, as noted above; family history, in fact, may mainly mark *ancillary* modifying genetic variation.

240   *K. M. Weiss* et al.

This is a major pandemic, with lethal consequences on a continental or even hemispheric scale. The hypothesis that these diseases are multiple reflections of a common underlying genetic susceptibility has not yet been tested. If the hypothesis is correct, then the answer must be found in some genetic difference(s) between Amerindians and other racial groups. It is likely that specific environmental risk factors can be identified definitively only after such genes have been identified. However, the pathogenic effects of modernisation are clear, and risk-reduction by dietary restriction, increased physical activity and the like are possible preventive measures. In Amerindian and admixed groups, the conditions we have classified as belonging to a 'syndrome' constitute the greatest current health problem and are a major cause of death. Solving this problem will have major benefits on a continental scale.

Acknowledgements

Support for work reported here was provided in part by the American Diabetes Association and the American Heart Association, and by the Wenner-Greu Foundation for Anthropological Research.

References

Baker, P. T. (1984). Migrations, genetics, and the degenerative diseases of South Pacific Islanders. In *Migration and Mobility* (ed. A. J. Boyce), pp. 209–39. London: Taylor & Francis.
Blangero, J. & Konigsberg, L. W. (1991). Multivariate segregation analysis using the mixed model. *Genetic Epidemiology* 8, 299–316.
Blangero, J., MacCluer, J. W., Kammerer, C. M., Mott, G. E., Dyer, T. D. & McGill, H. C. (1990). Genetic analysis of apolipoprotein A-I in two dietary environments. *American Journal of Human Genetics* 47, 414–28.
Bonney, G. E. (1986). Regressive logistic models for familial disease and other binary traits. *Biometrics* 42, 611–25.
Bonney, G. E. (1988). On the statistical determination of major gene mechanisms in continuous human traits: regressive models. *American Journal of Medical Genetics* 18, 731–49.
Bonney, G. E. & Elston, R. C. (1985a). Integrals of multinormal mixtures. *Applied Mathematics and Computation* 16, 93–104.
Bonney, G. E. & Elston, R. C. (1985b). Likelihood models for multivariate traits in human genetics. *Biometrical Journal* 5, 553–63.
Bonney, G. E., Lathrop, G. M. & Lalouel, J. M. (1988). Combined linkage and segregation analysis using regressive models. *American Journal of Human Genetics* 43, 29–37.
Brand, J. C., Snow, B. J., Nabhan, G. P. & Truswell, A. S. (1990). Plasma glucose and insulin responses to traditional Pima Indian meals. *American Journal of Clinical Nutrition* 51, 416–20.
Chakraborty, R. (1986). Gene admixture in human populations: models and predictions. *Yearbook of Physical Anthropology* 29, 1–43.

Chakraborty, R. & Weiss, K. (1986). Frequencies of complex diseases in hybrid populations. *American Journal of Physical Anthropology* **70**, 489–503.

Cohen, M. M., Young, T. K. & Hammarstrand, K. M. (1989). Ethnic variation in cholecystectomy rates and outcomes, Manitoba, Canada, 1972–84. *America Journal of Public Health* **79**, 751–5.

Diehl, A. K. & Stern, J. P. (1989). Special health problems of Mexican-Americans: obesity, gallbladder disease, diabetes mellitus and cardiovascular disease. *Advances in Internal Medicine* **34**, 73–96.

Elston, R. C. (1980). Segregation analysis. In *Current Developments in Anthropological Genetics*, vol. 1 (*Theory and Methods.*) (ed. J. H. Mielke & M. H. Crawford), pp. 327–54. New York: Plenum.

Elston, R. C. (1981). Segregation analysis. *Advances in Human Genetics* **11**, 63–120.

Elston, R. C. (1986). Modern methods of segregation analysis. In *Modern Statistical Methods in Chronic Disease Epidemiology* (ed. S. H. Moolgavkar & R. L. Prentice), pp. 213–24. New York: Wilen.

Evers, S., McCracken, E., Antone, I. & Deagle, G. (1987). The prevalence of diabetes in Indians and Caucasians living in southwestern Ontario. *Canadian Journal of Public Health* **78**, 240–3.

Haffner, S. M. (1987). Hyperinsulinemia as a possible etiology for the high prevalance of non-insulin-dependent diabetes in Mexican Americans. *Diabete et Metabolisme* **13**, 337–44.

Haffner, S. M., Diehl, A. K., Stern, M. P. & Hazuda, H. P. (1989). Central adiposity and gallbladder disease in Mexican Americans. *American Journal of Epidemiology* **129**, 587–95.

Haffner, S. M., Stern, M. P., Hazuda, H. P., Pugh, J., Patterson, J. K. & Malina, R. (1986a). Upper body and centralized adiposity in Mexican Americans and non-Hispanic Whites: relationship to body mass index and other behavioral and demographic variables. *International Journal of Obesity* **10**, 493–502.

Haffner, S. M., Stern, M. P., Hazuda, H. P., Rosenthal, M. & Knapp, J. A. (1986b). The role of behavioral variable and fat patterning in explaining ethnic differences in serum lipids and lipoproteins. *American Journal of Epidemiology* **123**, 830–9.

Haffner, S. M., Stern, M. P., Hazuda, H. P., Rosenthal, M., Knapp, J. A. & Malina, R. M. (1986c). Role of obesity and fat distribution in non-insulin-dependent diabetes mellitus in Mexican Americans and non-Hispanic Whites. *Diabetes Care* **9**, 153–61.

Hamman, R. F., Marshall, J. A., Baxter, J., Kahn, L. B., Mayer, E. J., Orleans, M., Murphy, J. R. & Lezotte, D. C. (1989). Methods and prevalence of non-insulin-dependent diabetes mellitus in a biethnic Colorado population. *American Journal of Epidemiology* **129**, 295–311.

Hanis, C. L., Ferrell, R. E., Barton, S. A., Aguilar, L., Garza-Ibarra, A., Tulloch, B. R., Garcia, C. A. & Schull, W. J. (1983). Diabetes among Mexican Americans in Starr County, Texas. *American Journal of Epidemiology* **118**, 659–72.

Hanis, C. L., Ferrell, F. E., Tulloch, B. R. & Schull, W. J. (1985). Gallbladder disease epidemiology in Mexican Americans in Starr County, Texas. *American Journal of Epidemiology* **122**, 820–9.

Hanis, C. L., Hewett-Emmett, D., Bertin, T. K. & Schull, W. J. (1991). The origins of U.S. Hispanics: implications for diabetes. *Diabetes Care* **14**, 618–27.

Howard, B. V. (1986). Obesity, cholelithiasis, and lipoprotein metabolism in Man.

242   K. M. Weiss et al.

In *Bile Acids and Atherosclerosis*, vol. 15. (ed. S. M. Grundy), pp. 169–86. New York: Raven Press.

Howard, B. V. (1987). Lipoprotein metabolism in diabetes mellitus. *Journal of Lipid Research* **28**, 613–28.

Iyengar, S., Hamman, R. F., Marshall, J. A., Baxter, J., Majumder, P. P. & Ferrell, R. E. (1989). Genetic studies of type 2 (non-insulin-dependent) diabetes mellitus: lack of an association with seven genetic markers. *Diabetologia* **32**, 690–3.

Knowler, W. C., Carraher, M. J., Pettitt, D. J. & Bennett, P. H. (1984). Epidemiology of cholelithiasis in the Pima Indians. In *Epidemiology and Prevention of Gallstone Disease* (ed. L. Capocaccia, G. Ricci, F. Angelico, M. Angelico & A. F. Attili), pp. 15–22. Hingham, Massachusetts: MTP Press.

Knowler, W. C., Pettitt, D. J., Bennett, P. H. & Williams, R. C. (1983). Diabetes mellitus in the Pima Indians: genetic and evolutionary considerations. *American Journal of Physical Anthropology* **62**, 107–14.

Knowler, W. C., Pettitt, D. J., Lillioja, S. & Nelson, R. G. (1988). Genetic and environmental factors in the development of diabetes mellitus in Pima Indians. In *Genetic Susceptibility to Environmental Factors – A Challenge for Public Intervention* (ed. U. Smith, S. Eriksson & F. Lindgarde), pp. 67–74. Stockholm: Almqvist & Wiksell International.

Knowler, W. C., Pettitt, D. J., Savage, P. J. *et al.* (1990). Diabetes mellitus in the Pima Indians: incidence, risk factors and pathogenesis. *Diabetes/Metabolism Reviews* **6**.

Laasko, M., Suhonen, M., Julkunen, R. & Pyorala, K. (1990). Plasma insulin, serum lipids and lipoproteins in gallstone disease in non-insulin dependent diabetic subjects: a case control study. *Gut* **31**, 344–7.

Lalouel, J. M. (1983). Segregation analysis of familial data. In *Methods in Genetic Epidemiology* (ed. N. E. Morton, D. C. Rao & J. M. Lalouel), pp. 75–97. Basle: S. Karger.

Lalouel, J. M., LeMignon, L., Simon, M., Fauchet, R., Bourel, M., Rao, D. C. & Morton, N. E. (1985). A combined qualitative (disease) and quantitative (serum iron) genetic analysis of idiopathic hemochromatosis. *American Journal of Human Genetics* **37**, 700–18.

Lalouel, J. M., Rao, D. C., Morton, N. E. & Elston, R. C. (1983). A unified model for complex segregation analysis. *American Journal of Human Genetics* **35**, 816–26.

Lange, K. & Boehnke, M. (1983). Extensions to pedigree analysis. IV. Covariance components models for multivariate traits. *American Journal of Medical Genetics* **14**, 513–24.

Lange, K. & Elston, R. C. (1975). Extensions to pedigree analysis. I. Likelihood calculations for simple and complex pedigrees. *Human Heredity* **25**, 95–105.

Lee, E. T., Anderson, P. S., Bryan, J., Bahr, C., Coniglione, T. & Cleves, M. (1985). Diabetes, parental diabetes, and obesity in Oklahoma Indians. *Diabetes Care* **8**, 107–13.

Lillioja, S., Mott, D. M., Zawakzki, J. K., Young, A. A., Abott, W. G. H., Knowler, W. C., Bennett, P. H., Mott, P. & Bogardus, C. (1987). In vivo insulin action is a familial characteristic in nondiabetic Pima Indians. *Diabetes* **36**, 1329–35.

Lowenfels, A. B., Lindstron, C. G., Conway, M. J. & Hastings, P. R. (1985). Gallstones and risk of gallbladder cancer. *Journal of the National Cancer Institue* **75**, 77–80.

Mueller, W. H., Joos, S. K., Hanis, C. L., Zavaleta, A. N., Eichner, J. & Schull, W. J. (1984). The Diabetes Alert Study: growth, fatness, and fat patterning, adolescence through adulthood in Mexican Americans. *American Journal of Physical Anthropology* **64**, 389–99.

Pettitt, D. J. & Knowler, W. C. (1988). Diabetes and obesity in the Pima Indians: a cross-generational vicious cycle. *Journal of Obesity and Weight Regulation* **7**, 61–75.

Raboudi, S. H., Mitchell, B. D., Stern, M. P., Eifler, C. W., Haffner, S. M., Hazuda, H. P. & Frazier, M. L. (1989). Type II diabetes mellitus and polymorphism of insulin-receptor gene in Mexican Americans. *Diabetes* **38**, 975–80.

Samet, J. M., Coultas, D. B., Howard, C. A., Skipper, B. J. & Hanis, C. L. (1988). Diabetes, gallbladder disease, obesity and hypertension among Hispanics in New Mexico. *American Journal of Epidemiology* **128**, 1302–11.

Stern, M. P. (1988). Type II Diabetes Mellitus: Interface between clinical and epidemiological investigation. *Diabetes Care* **11**, 119–25.

Stern, M. P. & Haffner, S. M. (1986). Body fat distribution and hyperinsulinemia as risk factors for diabetes and cardiovascular disease. *Arteriosclerosis* **6**, 123–30.

Szathmary, E. J. E. (1987). Genetic and environmental risk factors. In *Diabetes in the Canadian Native Population: Bio-Cultural Perspectives* (ed. T. K. Young), pp. 27–66. Toronto: Canadian Diabetes Association.

Szathmary, E. J. E. (1989). The impact of low carbohydrate consumption on glucose tolerance, insulin concentration and insulin response to glucose challenge in Dogrib Indians. *Medical Anthropology* **11**, 329–50.

Szathmary, E. J. E. (1990). Diabetes in Amerindian populations: the Dogrib studies. In *Health and Diseases of Populations in Transition* (ed. A. Swedlund & G. Armelagos), pp. 75–104. South Hadley, Massachusetts: Bergin and Garvey.

Trowell, H. C. & Burkitt, D. (eds) (1981). *Western Diseases*. London: Edward Arnold.

Valdez, R. (1991). A simple model-based index of abdominal adiposity. *Journal of Clinical Epidemiology* **44**, 955–9.

Weiss, K. M., Ferrell, R. E., Hanis, C. L. & Styne, P. N. (1984a). Genetics and epidemiology of gallbladder disease in New World native peoples. *American Journal of Human Genetics* **36**, 1259–78.

Weiss, K. M., Ferrell, R. E. & Hanis, C. L. (1984b). A New World Syndrome of metabolic diseases with a genetic and evolutionary basis. *Yearbook of Physical Anthropology* **27**, 153–78.

Weiss, K. M., Buchanan, A. V., Moore, J. M., Valdez, R., Ahn, Y. I., Russell, M. E., Campbell, J., Kris-Etherton, P. & Georges, E. (1993). Gene–environment interaction: diabetes and related diseases in Amerindians. In *Health and the Human Condition* (ed. P. Gindhart). Washington, D.C.: Smithsonian Press (in press).

West, K. M. (1978a). Diabetes in American Indians. In *Advances in Metabolic Diseases*, vol. 9 (ed. R. Levine & R. Luft), pp. 29–48. New York: Academic Press.

West, K. M. (1978b). *Epidemiology of Diabetes and its Vascular Lesions*. New York: Elsevier.

Young, T. K. & Roche, B. A. (1990). Factors associated with clinical gallbladder disease in a Canadian Indian population. *Clinical Investigations in Medicine* **13**, 55–9.

16 Sex ratio determinants in Indian populations: studies at national, state and district levels

A. H. BITTLES, W. M. MASON, D. N. SINGARAYER,
S. SHREENIWAS AND M. SPINAR

Introduction

Although the human sex ratio at conception, i.e. the primary sex ratio, is not readily amenable to investigation, data on the karyotypes of spontaneous and induced abortions indicate that early in pregnancy there is a large surplus of males, estimated at between 123 and 130 males for every 100 females (Stevenson, 1959; Lee & Takano, 1970; Hassold et al., 1983). Selective attrition of male foetuses occurs throughout the gestational period and also is seen in excess numbers of male stillbirths, resulting in a sex ratio at birth (i.e. secondary sex ratio) in the Caucasian populations of Western countries of 105 to 106, with slightly lower values of between 102 and 104 for Blacks (Chahnazarian, 1988). However, considerable global variation has been reported in secondary sex ratios, ranging from less than 100 to over 110 (Visaria, 1967; James, 1987).

While the excess spontaneous loss of male foetuses must be considered a primarily biological phenomenon, related to the male phenotype but not significantly associated with the expression of X chromosome mutations in hemizygous males (Stevenson & Bobrow, 1967), in postnatal life both biological and social factors influence the tertiary sex ratio. By the third decade approximately equal numbers of males and females are observed in Western countries, the decline in proportion of males again being ascribed to their relatively greater biological frailty, in combination with higher exposure to fatal accidents and homicide (Macfarlane & Mugford, 1984). The biological advantage conferred on females continues throughout adult life, hence the greater female life expectancy.

The picture is different in Southern Asia where the natural biological advantages of females observed in Western countries are frequently negated by social discrimination. Until the introduction of punitive legislation in the early to mid-nineteenth century, female infanticide appears to have been commonplace in North and West India; in 1836–7, tertiary sex ratios ranging from 300 to 1967 were recorded at district level among the Jharejas, a Rajput caste living in Gujarat (Pakrasi & Sasmal, 1971). In fact, a male preponderance in the population has been observed among the totals returned by all censuses of India from the earliest series dating back to 1872. Analyses of this persistent pattern in the census data considered explanations including female infanticide, excess mortality among women during child-bearing years, systematic under-enumeration of females, and migration (Visaria, 1967; El-Badry, 1969). As it proved impossible to explain the aberrant sex ratios in terms of migration or female under-enumeration, differential mortality by sex was implicated as the driving force.

While female infanticide now is reputedly rare in India, during the past decade reports have emerged of the establishment of clinics specialising in the determination of foetal sex and subsequent female foeticide (Jeffery *et al.*, 1984). The prevalence of these reports, mostly anecdotal in nature and with little or no supporting evidence, has provoked the Government of India to introduce legislation aimed at strictly limiting amniocentesis for the purpose of foetal sex determination, except where clinically warranted. That female children in Southern Asia are selectively disadvantaged both in terms of nutrition and health care is well documented (Chen *et al.*, 1981; Miller, 1981; Mosley & Chen, 1984; Sandhya, 1986), as are claims of higher female than male infant mortality, commencing in the post-neonatal period (Simmons *et al.*, 1982; Das Gupta, 1990). Female mortality remains high in adulthood; in India, deaths of mothers in childbirth are estimated at 500 per 100 000 live births (UNICEF, 1989). To date, female life expectancy at birth remains lower than that of males, although life expectancy at age 65 now shows the female advantage observed in Western societies (Martin, 1988, 1990).

Where the secondary and, more especially, the tertiary sex ratio is high there are grounds for supposing the existence of son preference. In the patriarchal and largely agrarian societies of Southern Asia, a greater value is placed on male children in the cultural and socioeconomic domains, which in turn leads to preferential treatment with respect to food and health care. Under such regimes, female children are more likely to be considered liabilities and be neglected, and thereby suffer elevated mortality rates (Houska, 1981; Miller, 1981; Minturn, 1984; Koenig & D'Souza, 1986; Das Gupta, 1987). However, the entire region, and particularly India, is far from homogeneous in terms of cultural values and socioeconomic organisation. South India, with its distinct kinship organisation and comparatively non-Brahmanic cultural system, is more gender-egalitarian than North India. Thus it has been suggested that, in combination with the important role accorded to women in rice cultivation, a greater degree of gender equality has resulted to the benefit of female children in South India (Miller, 1981; Dyson & Moore, 1983).

Non-economic factors that reinforce son preference have also been claimed to influence sex ratios. Among groups with strong son preference, once family fertility attains a target goal for number of males no further births are welcome, with families in practice exhibiting strong bias against higher-order females (Pakrasi & Halder, 1973; Das, 1984; Das Gupta, 1987). This pattern is to a degree enforced by the norm that a family marries all its daughters before arranging the marriages of sons. Birth of a daughter late in the mother's reproductive span could either seriously delay the marriage of first- or second-born sons, or provoke the culturally undesirable step of marrying a son before all daughters have married (Caldwell *et al.*, 1983).

The present study first examines both historical and contemporary statistics on the tertiary sex ratio in India at the state level. This evidence is used to provide tangible support for the preceding summary, and also to place in context the predominantly Hindu and agrarian South Indian state of Karnataka, which is the site of major ongoing studies into the effects of consanguineous marriage on fertility and mortality (Bittles *et al.*, 1990, 1991). The possibility and extent of gender bias in Karnataka is then considered, through more detailed examination of secondary and tertiary sex ratios based on census data for that state. If there is no gender preference in Karnataka, or if gender preference is undifferentiated by social factors, the secondary and tertiary sex ratios should not be skewed in favour of males, conditional on a variety of social characteristics measured by the census. Conversely, if gender preference does exist in Karnataka, this should be detectable through examination of sex ratios conditional on these social characteristics. Of particular relevance to the theme of this symposium will be the comparison of sex ratios between urban and rural localities. The value of this type of study, which complements those based on anthropological or sample survey data, is that it is based on population counts. This is helpful when studying the small differences in sex ratios that could provide evidence of this sometimes subtle phenomenon.

Data and methods

The data used for this study are taken from the Government of India (1984*a–c*) publications of the Census of India, 1981. The volumes on Karnataka (Government of India, 1984*c*) contain tables of children born and children surviving as reported by mothers, classified by sex and by selected characteristics of the mother: age, marital status (ever-married versus currently married), religion, education, occupation, labour force participation, place of residence (urban versus rural), and parity.

The published census data allow the calculation of sex ratios at birth and for surviving children, conditional on age and residence, while controlling one at a time for education, occupation-labour force participation, and religion. Although parity is included in the tables, it is recorded in insufficient detail for use in the analysis of sex ratios. Moreover, it is not possible to extract sex ratios simultaneously on all of the socioeconomic factors. Hence, sex ratios derived from the reports of women married at the time of the census are presented separately for children ever born and surviving children, conditional on residence and age; and considered individually, education, occupation-labour force participation, and religion. Corresponding estimated sex ratios based on logistic regressions are also presented. These estimates can be thought of as 'smoothed' in the statistical sense. Since an additive logistic model fits the data well in each instance, attention is focused primarily on the estimated sex ratios rather than the observed values.

Results

The geographical location of each Indian state is shown in Fig. 16.1. Table 16.1 presents data on the sex ratios of the general population by state, derived from decennial census returns from 1901 to 1981. Substantial state boundary changes occurred during this time period, and the pre-Independence figures represent composite returns from colonial India and the partially self-governing princely states. As a result, the meaning of the spatial variability in sex ratios over time is partially obscured. However, it is apparent that large variations in tertiary sex ratios existed between states, with a general, national trend through time towards more equal ratios. The only state with a majority of females throughout the period was Kerala, located in the southwest of the country, which in 1981 had much the highest rates of male (74.0%) and female (64.5%) literacy in India. Although Karnataka, also in South India, has had somewhat higher tertiary sex ratios than Kerala, it remains at the lower end of the sex ratio distribution by states.

The tertiary sex ratios by state for urban and rural residents in 1981 are shown in Table 16.2. With the exception of Tripura, all states have higher urban than rural ratios, probably owing to the movement of males into urban centres in search of improved employment prospects. In both rural and urban settings, the tertiary sex ratios for Karnataka are again among the lowest reported for India.

Table 16.3, which is based largely on data collated by UNICEF in 1987,

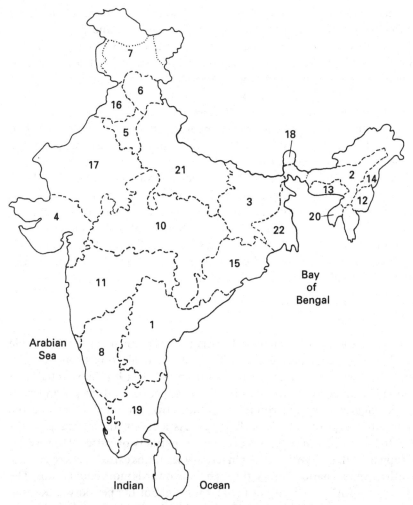

Fig. 16.1 Geographical location of the states of India, 1981: 1, Andhra Pradesh; 2, Assam; 3, Bihar; 4, Gujarat; 5, Haryana; 6, Himachal Pradesh; 7, Jammu and Kashmir; 8, Karnataka; 9, Kerala; 10, Madhya Pradesh; 11, Maharashtra; 12, Manipur; 13, Meghalaya; 14, Nagaland; 15, Orissa; 16, Punjab; 17, Rajasthan; 18, Sikkim; 19, Tamil Nadu; 20, Tripura; 21, Uttar Pradesh; 22, West Bengal.

presents information on sociodemographic variables for India as a whole. The high fertility of the population and low proportion of births attended by trained health personnel, very large numbers below the poverty line, especially in urban areas, poor access to potable water, low rates of literacy, high proportion of infants with low birth weight and childhood malnutrition, and low immunisation rates, all are factors that could determine

Table 16.1. *Tertiary sex ratio by state, 1901–81*

Number	State	1901	1911	1921	1931	1941	1951	1961	1971	1981
1	Andhra Pradesh	101.6	100.9	100.7	101.4	102.0	101.5	101.9	102.4	102.5
2	Assam	NA	NA	NA	NA	NA	NA	NA	NA	111.0
3	Bihar	94.8	95.8	98.5	100.7	100.4	101.0	100.6	104.9	105.7
4	Gujarat	104.8	105.7	105.9	105.8	106.3	105.1	106.3	107.0	106.2
5	Haryana	115.4	118.3	118.5	118.5	115.1	114.8	115.2	115.4	114.9
6	Himachal Pradesh	113.1	112.5	112.4	111.5	112.4	109.6	106.6	104.3	102.8
7	Jammu and Kashmir	113.4	114.2	114.9	115.6	115.1	114.5	114.0	113.9	112.1
8	Karnataka	101.7	102.0	103.2	103.6	104.2	103.5	104.3	104.5	103.9
9	Kerala	99.6	99.2	98.9	97.9	97.4	97.3	97.9	98.4	96.9
10	Madhya Pradesh	101.0	101.5	102.7	102.8	103.1	103.4	105.0	106.2	106.3
11	Maharashtra	102.2	103.5	105.3	105.6	105.4	106.3	106.8	107.5	106.7
12	Manipur	96.4	97.2	96.0	93.9	94.8	96.5	98.5	102.0	103.0
13	Meghalaya	NA	NA	NA	NA	NA	NA	NA	NA	104.9
14	Nagaland	102.8	100.8	100.9	100.3	97.9	100.1	107.2	114.9	115.9
15	Orissa	96.4	94.7	92.1	93.7	95.0	97.8	99.9	101.3	101.9
16	Punjab	120.3	128.2	125.2	122.8	119.6	118.5	117.2	115.6	113.8
17	Rajasthan	110.5	110.1	111.6	110.3	110.4	108.6	110.1	109.8	108.8
18	Sikkim	109.1	105.1	103.1	103.4	108.7	110.2	110.7	115.9	119.8
19	Tamil Nadu	95.8	96.0	97.2	97.3	98.8	99.3	100.8	102.2	102.4
20	Tripura	114.4	113.0	113.0	113.0	112.9	110.6	107.4	106.1	105.7
21	Uttar Pradesh	106.7	109.3	110.1	110.7	110.2	109.9	110.0	113.8	113.0
22	West Bengal	105.8	108.1	110.5	112.3	117.4	115.7	113.9	112.3	109.8
	All India	104.8	104.8	105.8	106.4	105.7	105.1	106.2	107.5	106.9
	Populations (millions)	283.9	303.1	305.8	338.2	314.9	357.0	435.7	548.2	683.1

NA, data not available.
Data based on Mitchell (1982) and Government of India (1984*a–c*).

survival in childhood and adulthood, and hence influence the tertiary sex ratio. The particular importance of maternal immunisation against the causative bacterium of tetanus (*Clostridium tetani*) is illustrated by its diagnosis in 52.4% of all neonatal deaths in a study from the North Indian state of Uttar Pradesh (Simmons *et al.*, 1982). The high mortality due to tetanus may be at least partially explicable in terms of the type of instrument used to cut the umbilical cord. Significantly more deaths occur when a sickle, as opposed to scissors or a blade, is used (Sandhya, 1986).

Tables 16.4–16.7 refer specifically to Karnataka. Owing to the tabulation procedures of the Indian Census, the total number of mothers from whose reports the sex ratios are derived varies across tables. In countries other than India, paternal age and parity have been shown to be negatively correlated with the secondary sex ratio (Curtsinger *et al.*, 1983; Khoury *et al.*, 1984; James & Rostron, 1985; Ruder, 1985, 1986) and there is good reason to expect the same influences to be operative in Karnataka.

Table 16.2. *Tertiary sex ratio by state, urban and rural, 1981*

		Tertiary sex ratio		Population (millions)	
Number	State	urban	rural	urban	rural
1	Andhra Pradesh	105.5	101.6	12.48	41.06
2	Assam	130.2	109.1	2.05	17.85
3	Bihar	120.2	103.8	8.72	61.20
4	Gujarat	110.5	104.3	10.60	23.48
5	Haryana	117.8	114.1	2.83	10.10
6	Himachal Pradesh	125.8	101.1	0.33	3.95
7	Jammu and Kashmir	114.3	111.5	1.26	4.73
8	Karnataka	108.0	102.3	10.72	26.41
9	Kerala	97.9	96.7	4.77	20.68
10	Madhya Pradesh	113.2	104.6	10.59	41.59
11	Maharashtra	117.6	101.3	21.99	40.79
12	Manipur	103.2	102.9	0.38	1.05
13	Meghalaya	110.6	103.6	0.24	1.09
14	Nagaland	145.2	111.2	0.12	0.65
15	Orissa	116.5	100.1	3.11	23.26
16	Punjab	115.7	113.1	4.65	12.14
17	Rajasthan	114.0	107.5	7.21	27.05
18	Sikkim	143.4	115.8	0.05	0.27
19	Tamil Nadu	104.5	101.3	15.95	32.46
20	Tripura	104.5	105.8	0.23	1.83
21	Uttar Pradesh	118.2	111.9	19.90	90.96
22	West Bengal	122.1	105.6	14.45	40.13

Data based on Government of India (1984*a–c*).

Unfortunately the census data do not allow control for these variables. Therefore, in Tables 16.4–16.7, mother's age at the time of the Census is used as a proxy for both paternal age and parity. By this reasoning, if male preference exists it should be manifest in higher sex ratios among the children of older mothers. Each of these tables also presents sex ratios separately for urban and rural mothers. For the secondary sex ratio, male preference should lead to higher ratios in rural settings. For the tertiary sex ratio, male preference should accentuate this pattern. Selective migration of males from rural to urban areas should not obscure it, since the sex ratios are based on mothers' reports irrespective of whether their children may have migrated.

Table 16.4 presents secondary and tertiary sex ratios conditional on mother's residence and age. For this table only, the fitted sex ratios are based on an additive logistic regression in which the regressors are residence and age (and no additional variable), treated as nominal dimensions. The table shows that secondary sex ratios are higher in

Table 16.3. *Parameters influencing premature mortality in India*

Demographic profile	
Population	802.1 million
Annual number of births	26.8 million
Population < 5 years of age	111.5 million
Population < 16 years of age	315.4 million
Crude birth rate	32 / 1000 live births
Crude death rate	11 / 1000
Life expectancy at birth	59 years
Annual population growth rate (1980–86)	2.2%
Socioeconomic background	
GNP per capita (1986)	U.S.$290
Daily per capita calorie intake as % of requirements (1985)	94%
Household income expenditure on food (1980–85)	52%
Population below poverty line (1977–86): urban	40%
rural	51%
Central government expenditure (1986): health	2.1%
education	2.1%
Population with access to drinking water: urban	80%
rural	47%
Adult literacy: males	57%
females	29%
Maternal factors	
Contraceptive prevalence (1981–85)	34%
Total fertility rate	4.2
Births attended by trained health personnel	33%
Pregnant women immunised against tetanus (1986–87)	47%
Maternal mortality rate (1980–87)	500 / 100 000 live births
Infant and childhood parameters	
Infants with low birth weight (1982–87)	30%
One year-old infants fully immunised: TB	46%
DPT	58%
polio	50%
measles	17%
Prevalence of wasting (12–23 months of age) (1980–87)	37%
Children < 5 years of age with mild/moderate malnutrition (1980–86)	33%
Children < 5 years of age with severe malnutrition (1980–86)	5%
Infant mortality < 1 year of age	100 / 1000 live births
Childhood mortality < 5 years of age	152 / 1000 live births
Infant and childhood deaths (< 5 years)	4.1 million

Except where otherwise stated all data refer to 1987.
Based on UNICEF (1989).

Table 16.4. *Maternal age and sex ratios of all liveborn and living children in urban and rural populations, Karnataka, 1981*

	Sex ratio: urban		Sex ratio: rural		
Age	liveborn	living	liveborn	living	Numbers of mothers
15–19	106.7	104.8	105.3	103.9	644 468
	(105.9)[a]	(104.0)	(105.6)	(104.1)	
20–24	106.0	105.3	105.6	104.3	1 270 307
	(105.9)	(104.5)	(105.6)	(104.6)	
25–29	105.9	105.5	104.9	104.1	1 360 995
	(105.4)	(104.5)	(105.1)	(104.6)	
30–34	105.9	104.9	105.6	105.4	1 059 216
	(105.4)	(105.2)	(105.6)	(105.3)	
35–39	106.0	105.4	106.9	107.2	947 708
	(106.9)	(106.6)	(106.5)	(106.7)	
40–45	107.9	108.1	107.9	109.1	729 748
	(108.1)	(108.7)	(107.8)	(108.9)	
45–49	110.6	111.7	108.9	110.7	556 459
	(110.0)	(110.9)	(109.2)	(111.0)	
Total number studied					6 568 901

[a] Numbers in parentheses are estimated sex ratios (fitted values) determined from an additive logistic regression in which the regressors are residence and age, each treated as discrete dimensions.
Source: Tables F17 (at birth) and F24 (surviving) in the Karnataka Census.

Karnataka than in Western populations. This confirms the results of a prospective, hospital-based study conducted in the state (Bittles *et al.*, 1988) and suggests that, in general, the local tertiary sex ratios also may be higher to an equivalent extent. Both secondary and tertiary sex ratios are greater in the progeny of women aged 35 years and older but, net of age, the sex ratios based on the reports of urban and rural women are approximately the same. There is an interaction involving type of sex ratio and age. The tertiary sex ratio is lower than the secondary sex ratio for younger women, but the difference erodes and then reverses with increasing age.

The overall literacy rates in Karnataka were 48.6% for males and 27.8% for females. If son preference abates as populations increase in educational status, and non-agrarian opportunities become available, mother's education should be inversely related to both the secondary and the tertiary sex ratio. Table 16.5 allows consideration of this possibility. Contrary to the results of other Southern Asia studies (Cleland & van Ginneken, 1989), an inverse relationship is indeed apparent between sex ratio and educational

Table 16.5. *Maternal education and sex ratios of liveborn and living children in urban and rural populations, Karnataka, 1981*

Education	Sex ratio: urban		Sex ratio: rural		Numbers of mothers
	liveborn	living	liveborn	living	
Illiterate	107.4	107.4	106.7	107.1	4 749 802
	(107.3)[a]	(107.4)	(106.7)	(107.1)	
Literate	107.1	106.5	106.7	106.0	941 244
	(107.3)	(106.6)	(106.6)	(106.0)	
Middle	106.3	105.4	104.3	103.3	463 995
	(105.8)	(104.9)	(105.0)	(104.1)	
Matriculate	105.3	104.6	107.5	106.6	352 597
	(106.0)	(105.3)	(105.1)	(104.4)	
Graduate	103.0	102.8	105.7	107.5	61 258
	(103.3)	(103.3)	(102.5)	(102.5)	
Total number studied					6 568 896

[a] Numbers in parentheses are estimated sex ratios (fitted values) determined from an additive logistic regression in which the regressors are maternal education, residence, and age, all treated as discrete dimensions.
Source: Tables F14 (at birth) and F24 (surviving) in the Karnataka Census.

attainment, seen most clearly for graduates versus those of lower attainment. Controlling for education suggests slightly higher values in urban centres in Karnataka. There also is the suggestion of a difference between secondary and tertiary sex ratios, especially among urban women for whom the secondary sex ratios are slightly higher, at educational levels below graduate status.

Sex ratios conditional on labour force participation are given in Table 16.6. Occupation, as recorded in the census tables (essentially a non-manual, manual, agricultural, non-agricultural classification), proved to be of little importance and so the results are not presented. However, the sex ratios do vary by labour force status, with mothers not in the labour force showing the highest sex ratios. In this instance, there are no differences between the sex ratios for children ever born or surviving; controlling for labour force status, the sex ratios of children born to rural mothers appear to be higher. Mothers not in the labour force are more likely than mothers who report themselves as employed to be traditional in terms of gender-role. Moreover, this category is likely to include a substantial percentage of unpaid family workers. Thus, despite the small differences across categories, the pattern is consistent with male preference.

Table 16.6. *Maternal labour force status and sex ratios of all liveborn and living children in urban and rural populations, Karnataka, 1981*

	Sex ratio: urban		Sex ratio: rural		
Labour force status	liveborn	living	liveborn	living	Numbers of mothers
Full-time	103.6	103.1	105.0	105.3	2 008 607
	(104.4)[a]	(104.2)	(104.8)	(105.1)	
Part-time	103.0	103.2	106.5	106.7	730 564
	(105.9)	(105.6)	(106.4)	(106.5)	
Not in labour force	107.7	107.3	107.9	108.0	3 835 967
	(107.5)	(107.1)	(108.1)	(108.2)	
Total number studied					6 575 138

[a] Numbers in parentheses are estimated sex ratios (fitted values) determined from an additive logistic regression in which the regressors are labour force status, residence, and age, all treated as discrete dimensions.
Source: Table F19 in the Karnataka Census.

Sex ratios conditional on religion are presented in Table 16.7. The major religious groups in terms of size are Hindu, Muslim and Christian, respectively. Secondary and tertiary sex ratios are lowest among Christians, next lowest among Hindus, followed by Muslims, and highest among the heterogeneous 'Other' group, which is numerically small and includes Buddhists, Jains, Sikhs, Parsis and Jews. Controlling religion, there is little difference between the patterns of secondary and tertiary sex ratios, but there does appear to be a small interaction between residence and sex ratio. For the secondary sex ratio, the urban values are slightly higher than rural whereas, for surviving children, the sex ratios reported by urban mothers are lower than those of their rural counterparts.

To summarize, the main findings in Karnataka indicate that social factors influence both secondary and tertiary sex ratios. Maternal age exerts a limited effect, with older, currently married women reporting a greater excess of male progeny, at birth and among their surviving children. Greater educational attainment is associated with lower sex ratios, again, for both secondary and tertiary ratios. Relative to full-time and part-time workers, women not in the labour force report a preponderance of male progeny, at birth and among survivors. There are differences by religion; of particular significance with respect to the theme of urbanisation, residence (urban versus rural) at most exerts a modest effect on the sex ratios of the population.

Table 16.7. *Maternal religion and sex ratios of liveborn and living children in urban and rural populations, Karnataka, 1981*

Religion	Sex ratio: urban		Sex ratio: rural		Numbers of mothers
	liveborn	living	liveborn	living	
Hindu	106.8	106.4	106.4	106.7	5 690 707
	(106.7)[a]	(106.3)	(106.4)	(106.7)	
Muslim	107.4	107.5	108.1	108.5	699 440
	(107.9)	(107.7)	(107.6)	(108.2)	
Christian	105.2	103.8	103.2	104.5	117 908
	(104.3)	(103.9)	(104.1)	(104.4)	
Others	110.0	109.4	108.4	107.8	79 246
	(108.4)	(107.6)	(109.4)	(109.1)	
Total number studied					6 587 301

[a] Numbers in parentheses are estimated sex ratios (fitted values) determined from an additive logistic regression in which the regressors are religion, residence, and age, all treated as discrete dimensions. The religion classification used in the logistic regression includes all available detail on religion (so that 'Others' are distinguished as individual religions).
Source: Tables F16 (at birth) and F23 (surviving) in the Karnataka Census.

Discussion

Although the unadjusted sex ratios for Karnataka reported in Tables 16.1 and 16.2 are relatively low, there is nevertheless evidence of systematic variability in the sex ratio by mother's age, education, labour force participation and religion (Tables 16.4–16.7). In this context, the fundamental lack of differences between urban and rural sex ratios (Table 16.4) is surprising. It may be that in Karnataka the urban settings are insufficiently industrialised and 'modern' for an urban–rural differentiation to emerge.

In a society where son preference is known to exist, the finding of variable anti-female bias in the tertiary sex ratios is not surprising, and behavioural explanations of the phenomenon are well known. By contrast, social and behavioural factors that affect secondary sex ratios are not so evident. Women of differing socioeconomic backgrounds may state the number of their children in non-comparable ways. In addition, older women may find it difficult to remember the exact number of liveborn children they have had, especially if they are rural, less educated, or from a lower-class or more traditional background. Such women might recall their more valued sons with greater accuracy than they would daughters. The Karnataka data are partially consistent with this interpretation. The higher secondary sex ratios found among the offspring of older women are also consistent with the pattern noted earlier, that older, high-parity women try to stop childbearing once they have the desired number of sons, which is greater than the number of daughters. Finally, it is also possible that the higher secondary sex ratios derived from the reports of the older women are indicative of earlier female infanticide. In all of this, it should be borne in mind that in most instances the sex ratios for Karnataka do not approach the extreme values observed in many North Indian states (Tables 16.1 and 16.2).

The analysis presented here is hampered in two important ways. Because of the manner in which the census tabulations were published it was not possible to simultaneously control age, residence, education, occupation-labour force participation, and religion. For example, it could not be determined whether there is a labour force participation effect controlling education or *vice versa*, or if there is an interaction between education and labour force participation in the determination of sex ratios. What we have found is that, controlling education, the sex ratios are higher for urban than for rural mothers. Controlling labour force participation, the sex ratios are higher for rural than for urban mothers. This inconsistency of the urban–rural differential across controls means that the present analysis cannot speak decisively as to the validity of a materialist interpretation of the value of women in South India.

A second problem concerns potentially important omitted variables, in particular marital consanguinity. In contrast with North India, where marriages between close relatives are rigorously avoided by Hindus (Kapadia, 1958), in Karnataka and the other southern states, consanguineous marriages on average account for 20–30% of all unions (Bittles *et al.*, 1987, 1991). A salient characteristic of inbred marriages is the reduced level of dowry or bride-wealth payable, on occasion with total exemption from payment (Govinda Reddy, 1988; Caldwell *et al.*, 1988). Under such a system, and with the transition to dowry in South India (Caldwell *et al.*, 1983), the financial burden held to be associated with daughters would be greatly reduced in consanguineous marriages, which in turn would be linked to lessened anti-female bias in infancy and the premarital years. As consanguineous marriages are more common in rural areas, and among the poorest and least educated sections of the population (Dronamraju & Meera Khan, 1963; Rao & Inbaraj, 1977), failure to control for consanguinity could disguise a more substantial urban–rural sex ratio differential. Information on this subject is not available from census records, and it is on this topic that our continuing investigations into sex ratio determinants in South India are centred, using other data.

Acknowledgements

Financial assistance to A.H.B. from The Stopes Research Fund and The Fulbright Commission, to D.N.S. from The Ruggles-Gates Award of King's College, London, and to M.S. from the Student Research Opportunity Program of the University of Michigan, is gratefully acknowledged.

References

Bittles, A. H., Mason, W. M., Greene, J. & Appaji Rao, N. (1991). Reproductive behavior and health in consanguineous marriages. *Science* **252**, 789–94.
Bittles, A. H., Radha Rama Devi, A. & Appaji Rao, N. (1988). Consanguinity, twinning and secondary sex ratio in the population of Karnataka, South India. *Annals of Human Biology* **15**, 455–60.
Bittles, A. H., Radha Rama Devi, A. & Appaji Rao, N. (1990). Inter-relationships between consanguinity, religion and fertility in Karnataka, South India. In *Fertility and Resources* (ed. J. Landers & V. Reynolds), pp. 62–75. Cambridge University Press.
Bittles, A. H., Radha Rama Devi, A., Savithri, H. S., Rajeshwari Sridhar & Appaji Rao, N. (1987). Consanguineous marriage and postnatal mortality in Karnataka, South India. *Man* (New Series) **22**, 736–45.
Caldwell, J. C., Reddy, P. H. & Caldwell, P. (1983). The causes of marriage change in South India. *Population Studies* **37**, 343–61.
Caldwell, J. C., Reddy, P. H. & Caldwell, P. (1988). *The Causes of Demographic Change: Experimental Research in South India.* Madison: University of Wisconsin Press.
Chahnazarian, A. (1988). Determinants of the sex ratio at birth: review of recent literature. *Social Biology* **35**, 214–35.
Chen, L. C., Huq, E. & D'Souza, S. (1981). Sex bias in the family allocation of food and health care in rural Bangladesh. *Population and Development Review* **7**, 55–70.
Cleland, J. & van Ginneken, J. (1989). Maternal schooling and childhood mortality. In *Health Interventions and Mortality Change in Developing Countries* (ed. A. G. Hill & D. F. Roberts), pp. 13–34. Cambridge: Parkes Foundation.
Curtsinger, J. W., Ito, R. & Hiraizumi, Y. (1983). A two-generational study of human sex-ratio variation. *American Journal of Human Genetics* **35**, 951–61.
Das, N. (1984). Sex preference pattern and its stability in India: 1970–1980. *Demography, India* **13**, 108–19.
Das Gupta, M. (1987). Selective discrimination against female children in rural Punjab, India. *Population and Development Review* **13**, 77–100.
Das Gupta, M. (1990). Death clustering, mother's education and the determinants of child mortality in rural Punjab, India. *Population Studies* **44**, 489–505.
Dronamraju, K. R. & Meera Khan, P. (1963). A study of Andhra marriages, consanguinity, caste, illiteracy and bridal age. *Acta Genetica et Statistica Medica* **13**, 21–9.
Dyson, T. & Moore, M. (1983). On kinship structure, female autonomy, and demographic behavior in India. *Population and Development Review* **9**, 35–59.

El-Badry, M. A. (1969). Higher female than male mortality in some countries of South Asia: a digest. *Journal of the American Statistical Association* **64**, 1234–44.

Government of India (1984*a*). *Census of India 1981*. Series 1. Paper 3. *Household Population by Religion of Head of Household* (ed. V. S. Verma). New Delhi: Government of India Text Book Press.

Government of India (1984*b*). *Census of India 1981*. Part IIA and B, Part II Special, Series 2, 4–8, 10–13, 15–23. New Delhi: Government of India Text Book Press.

Government of India (1984*c*). *Census of India 1981*. Series 9. *Karnataka*, Part II-A, *General Population Tables* (ed. B. K. Das). Mysore: Government of India Text Book Press.

Govinda Reddy, P. (1988). Consanguineous marriages and marriage payment: a study among three South Indian caste groups. *Annals of Human Biology* **15**, 263–8.

Hassold, T., Quillen, S. D. & Yamane, J. A. (1983). Sex ratio in spontaneous abortions. *Annals of Human Genetics* **47**, 39–47.

Houska, W. (1981). The characteristics of son preference in an urban scheduled caste community. *Eastern Anthropologist* **34**, 27–35.

James, W. H. (1987). The human sex ratio. Part 1: a review of the literature. *Human Biology* **59**, 721–52.

James, W. H. & Rostron, J. (1985). Parental age, parity and sex ratio in births in England and Wales, 1968–1977. *Journal of Biosocial Science* **17**, 47–56.

Jeffery, R., Jeffery, P. & Lyon, A. (1984). Female infanticide and amniocentesis. *Social Science and Medicine* **19**, 1207–12.

Kapadia, K. M. (1958). *Marriage and Family in India*, 2nd edn. Calcutta: Oxford University Press.

Khoury, M. J., Erickson, J. D. & James, L. M. (1984). Paternal effects on the human sex ratio at birth: evidence from inter-racial crosses. *American Journal of Human Genetics* **36**, 1103–11.

Koenig, M. A. & D'Souza, S. (1986). Sex differences in childhood mortality in rural Bangladesh. *Social Science and Medicine* **22**, 15–22.

Lee, S. & Takano, K. (1970). Sex ratio in human embryos obtained from induced abortion: histological examination of the gonad in 1,452 cases. *American Journal of Obstetrics and Gynecology* **108**, 1294–7.

Macfarlane, A. & Mugford, M. (1984). *Birth Counts: Statistics of Pregnancy and Childbirth*. London: HMSO.

Martin, L. (1988). The aging of Asia. *Journal of Gerontology* **43**, S99–S113.

Martin, L. G. (1990). The status of South Asia's growing elderly population. *Journal of Cross-Cultural Gerontology* **5**, 93–117.

Miller, B. D. (1981). *The Endangered Sex: Neglect of Female Children in Rural North India*. Ithaca: Cornell University Press.

Minturn, L. (1984). Changes in the differential treatment of Rajput girls in Khalapur: 1955–1975. *Medical Anthropology* **8**, 127–32.

Mitchell, B. R. (1982). *International Historical Statistics, Africa and Asia*, pp. 43–6. New York: New York University Press.

Mosley, H. W. & Chen, L. C. (1984). An analytical framework for the study of child survival in developing countries. *Population Development Review* **10** (suppl.), 25–45.

Pakrasi, K. & Halder, A. (1973). The sex ratio at birth in India by parity and region. *Acta Medica Auxologica* **5**, 41–55.

Pakrasi, K. & Sasmal, B. (1971). Infanticide and variation of sex-ratio in a caste population of India. *Acta Medica Auxologica* **3**, 217–28.

Rao, P. S. S. & Inbaraj, S. G. (1977). Inbreeding in Tamil Nadu in South India. *Social Biology* **24**, 281–8.

Ruder, A. (1985). Paternal-age and birth-order effect: the human secondary sex ratio. *American Journal of Human Genetics* **37**, 362–72.

Ruder, A. (1986). Paternal factors affect the sex ratio. *Human Biology* **58**, 357–66.

Sandhya, S. (1986). Socio-cultural and economic correlates of infant mortality: a case study of Andhra Pradesh. *Demography, India* **15**, 86–102.

Simmons, G. B., Smucker, C., Bernstein, S. & Jensen, E. (1982). Post neonatal mortality in rural India: implications of an economic model. *Demography* **19**, 371–89.

Stevenson, A. C. (1959). Observations on the results of pregnancies in women resident in Belfast. III. Sex ratio with reference to nuclear sexing of chorionic villi of abortions. *Annals of Human Genetics* **23**, 415–20.

Stevenson, A. C. & Bobrow, M. (1967). Determinants of sex proportions in man, with consideration of the evidence concerning a contribution from X-linked mutations to intrauterine death. *Journal of Medical Genetics* **4**, 190–221.

Visaria, P. M. (1967). Sex ratio at birth in territories with a relatively complete registration. *Eugenics Quarterly* **14**, 132–42.

UNICEF (1989). *The State of the World's Children 1989*, pp. 93–113. Oxford University Press.

17 *Polarisation and depolarisation in Africa*

J. I. CLARKE

The World context

One of the remarkable facts of the twentieth century has been the growing polarisation of population into smaller areas of the earth's surface, particularly towns and cities. Urban centres localise an ever-increasing proportion of the world's population despite its rapid growth. It is estimated by the UN (1989) that between 1975 and 2000 two thirds of all population growth will take place in urban areas, so that while 29% of the world's population in 1950 lived in urban areas, in 1990 about 43% are living in them and by 2000 it will probably be 47%.

More and more of the world's urban dwellers are found in less developed countries (LDCs), which were only 16% urban in 1950, but about 34% in 1990 and will probably be 40% urban by 2000. The annual urban population growth of LDCs has usually exceeded 3.5%, although recently it has declined slightly, while in the more developed countries (MDCs) it has declined markedly from 1.9% in the late 1960s to 0.8% in the late 1980s. Obviously, urban growth in the LDCs has far exceeded the capacity of countries to cope with it: hence the deteriorating conditions in overburdened towns and cities, many of which, like Mexico City, Sao Paulo, Calcutta, Bombay and Shanghai, are among the largest in the world.

Africa presents a particularly poignant example of the problems involved, as it has the fastest population and urban growth in the world as well as the lowest economic development and growth and many of the poorest countries, especially in Tropical Africa. Thus it exemplifies in stark reality many of the worst difficulties of urban health and ecology.

African urbanisation

Until recent decades, Africa's level of urban population was one of the lowest in the world, with probably only 5% urban in 1900 and 12% in 1950. However, since the middle of this century it has risen rapidly to reach about

34% in 1990, slightly above the LDC average (33%); it is expected to be over 41% by the end of the century. This remarkable rise is because Africa's annual rate of urban growth is probably the highest in the world: 5% in 1980–85 compared with 3.5% for LDCs as a whole and more than twice the annual growth rate of the rural population of Africa (2.2%).

Between 1900 and 1950 Africa's urban population grew by about 15 million, but during the 1950s alone it grew by 16 million, during the 1960s by 26 million, and during the 1970s by 52 million. Over the past decade the urban population has probably doubled, so that in 1990 roughly 220 million Africans are living in towns, compared with only about 27 million 40 years ago. By the year 2000 some 350 million may be living in African towns, so that in the course of half a century (1950–2000) the urban population will have multiplied more than twelve times. This veritable urban explosion is much more traumatic than ever experienced by the more developed countries, especially as it has not been accompanied by rapid economic development. There has been no industrial revolution in Africa to absorb the new labour-force and provide new wealth.

Of course, the picture is not everywhere the same throughout the continent. There are quite strong variations in the level of urban population, some of which may be attributed to the history of urbanisation and modern economic development (Fig. 17.1). In consequence, it is highest in northern Africa (where Libya is about 70% urban) and in southern Africa, and lower in eastern Africa (where Uganda, Rwanda and Burundi are little more than 10% urban) than in middle or western Africa. In all, a dozen countries are less than one quarter urban, most of which are in Tropical Africa, although we should remind ourselves that China also falls into this category. Generally also, coastal countries of Africa have higher proportions of urban population than the 14 land-locked countries (a reflection of greater colonial influence) although the land-locked Central African Republic and Zambia are notable exceptions, being well over 40% urban.

As for annual rates of urban growth, they tend to be inversely related to the level of urban population (Fig. 17.2), so that they are highest in feebly urbanised eastern Africa, where the average in 1980–85 was 6.8%, but over 10% in Tanzania and Mozambique, and lowest (3.7%) in more urbanised northern and southern Africa (UNECA, 1989) (Fig. 17.3).

Polarisation and primacy

A characteristic feature of African urbanisation is that it tends to be polarised in large cities. Africa does not rival Latin America in its number of mega-cities – only Greater Cairo is among the 25 cities in the world with

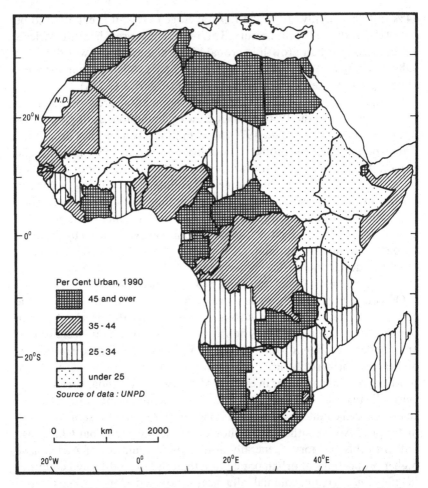

Fig. 17.1 The level of urbanisation in African countries, 1990. Northern, Southern and West-Central Africa tend to be much more urbanised than the rest of Tropical Africa.

more than 10 million inhabitants – but the continent now contains at least a dozen cities with more than 2 million inhabitants: Addis Ababa, Alexandria, Algiers, Cairo, Cape Town, Casablanca, Ibadan, Johannesburg, Kano, Greater Khartoum, Kinshasa and Lagos. Indeed, it has been forecast that there will be 29 such cities by the end of the century. Many large cities have been doubling in population size every ten years or so; Lagos, Kinshasa, Abidjan and Dar es Salaam are noteworthy examples, but lesser cities are also growing rapidly.

The urban systems of countries are generally dominated by the largest

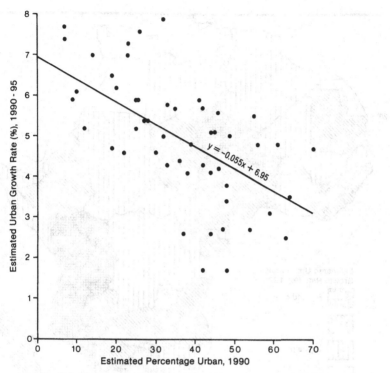

Fig. 17.2 The relation between levels and rates of urbanisation in African countries; this relation is weakly negative.

cities, which are very much larger than other cities and are often capitals. This urban primacy is so marked that primate cities may incorporate a tenth or more of the total population and have 5–10 times as many inhabitants as the second largest city; as in Angola, Benin, Central African Republic, Ivory Coast, Ethiopia, Madagascar, Malawi, Senegal, Sierra Leone, Tanzania, Togo, Tunisia and Uganda, where the largest cities have at least twice as many inhabitants as the next three cities in each country. Indeed, in some countries the capital may be the only major city. Only in a minority of countries is urban primacy weak, and log-normal or intermediate city-size distributions prevail (Clarke *et al.*, 1975), as in Cameroon, Egypt, Nigeria, South Africa, Zambia and Zimbabwe, where urban systems are more complex, for a variety of historical, political and/or economic reasons.

Polarisation is much more than demographic; it encompasses many other aspects of society, for primate cities concentrate administrations, industries, commercial activities, communications nodes, higher educational institutions, health facilities, entertainments, salary and wage

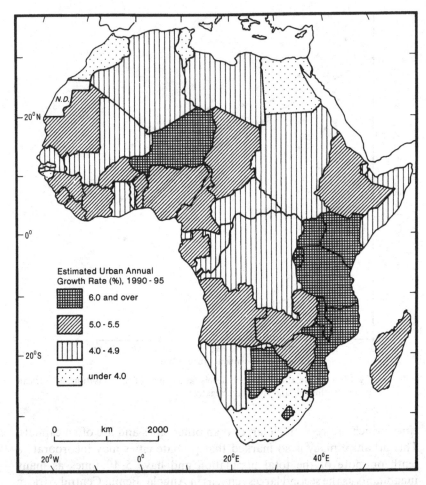

Fig. 17.3 Estimated urban annual growth rates of African countries, 1990–95. The rates are lower in Northern and Southern Africa than in much of Tropical Africa.

earners, and so on. The level of concentration obviously varies considerably, but it is not unusual for half of the industries, hospitals, banks, colleges and other activities to be localised within the primate city. This is of course partly a reflection of the fact that most African primate cities, like those in Latin America, are products of the colonial impact of the nineteenth and twentieth centuries, or greatly affected by that impact. The majority are city-ports or have peripheral locations as gateways that linked core regions with colonial powers. Focusing foreign presence and investments, they have accumulated an inordinate number of functions through cumulative causation. Unfortunately, their current population growth is in spite of slow economic growth or stagnation, and is often triggered by political insecurity, as in Kinshasa, or by economic disasters, which are all too frequent in Africa. With inadequate productive functions and infrastructures to absorb their burgeoning populations, primate cities localise many of the problems of spatial disharmony in their countries.

Urban primacy in fact reflects the excessive post-colonial political fragmentation of Africa, with 48 mainland countries and some 14 island states or dependencies, a dozen of which have less than a million inhabitants each. Urban systems are evolving within a complex jigsaw of states whose population–area relationships are extremely diverse, ranging from large and heterogeneous populations and territories, as in Ethiopia and Nigeria, to micro-states like Benin and Guinea Bissau. Very few cities are supranational in character, though Cairo and perhaps Addis Ababa may be seen in this light. Therefore urban hierarchies and spacing largely relate to individual state populations and territories, though the trend towards growing international migration, legal and illegal, may have an increasing effect, the wealthier countries attracting migrants from the poorer least developed countries (LLDCs) and land-locked countries.

Demographic urbanisation

The recent growth of towns and cities in Africa is closely linked with the early phases of demographic and mobility transitions, associated with rapid population growth and strong rural–urban migration. Africa has proceeded less along the paths of demographic and mobility transitions than other parts of the developing world, and its urbanisation has been termed demographic urbanisation rather than economic urbanisation (Escallier, 1988, p. 179), because it is not caused by radical transformations in agricultural productivity or industrialisation.

The main causes of urbanisation are the twin processes of natural increase and net migration to towns, the former being more continuous and the latter more volatile. Their relative significance varies in time and place, though net migration tends to be more important in the early stages of town growth, while the natural increase component becomes more significant later as younger adult migrants contribute to raising birth rates and lowering death rates. Often it is underestimated because of the massive influx of migrants, but it should be noted that African cities are located within the continent with the highest natural population increase of any world region, currently 2.9% per annum (or a doubling every 24 years), well above the LDC average (2.0%). In the 1990s, Africa is expected to have a natural increase rate of around 3% per annum, the fastest rate ever known for a world region. Only a handful of small island countries which

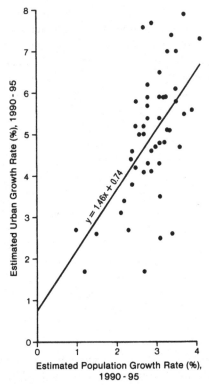

Fig. 17.4 The relation between the rates of urban and population growth in African countries; this relation is moderately positive.

have experienced fertility decline have rates below 2%: Mauritius, Reunion, Saint Helena, and the Seychelles. High natural increase results from sustained high fertility in comparison with only modest mortality decline, and it is only during the 1980s that many African countries have come to realise that their population growth is too rapid, and that it is positively related to urban growth (Fig. 17.4).

Urban fertility

Africa is demographically unusual in that during the 1980s it continued to experience high fertility, over 40 countries having birth rates of 40 per thousand or more and total fertility rates (TFRs) of 5 or more. In 1990 its average birth rate is 43 per thousand and average TFR 5.1, in comparison with the averages for LDCs of 29 and 3.7. Only a few major countries, like Egypt and Tunisia, have undergone some fertility decline, though Gabon has lower fertility for pathological reasons and South Africa has strong ethnic differential fertility (Fargues, 1988).

Unfortunately, urban–rural fertility data are not available for all African countries, so a generalised picture of differentials cannot be given, but in all those countries for which data are available it is evident that urban fertility is lower than rural fertility. This is particularly true for larger towns and for the North African countries of Egypt, Morocco and Tunisia as well as Kenya, where urban TFRs are more than 2 children lower than rural TFRs (Fargues, 1988, p. 192). Nevertheless, African urban TFRs are very rarely below 4, and more usually 5–7.

Caldwell's (1976) theory of intergenerational flows of wealth certainly helps to explain this incipient urban fertility decline, with wealth beginning to pass down the age groups instead of the opposite direction, though this is more apparent in North and South African cities than in Tropical Africa. There rural–urban links are much stronger, multi-local residence more common, and the contrasts between rural and urban ways of life much less marked. The extended family is still the economic foundation of life in Tropical African towns, and thus greatly influences urban fertility. For example, the incidence of early marriage of women and the high proportions married are not always lower in urban areas. In some countries, such as Cameroon, Ivory Coast, Nigeria and Zaïre, it seems that urban marriages are earlier than rural marriages, but contrary evidence is available from Benin, Ghana and Senegal (Fargues, 1988, p. 194). In addition, pre-marital fertility is quite high in Tropical Africa, a situation that is rare in North Africa, where women marry much later, especially in towns where their average age at marriage is now 23–24.

The desire for children is still strong among women in African towns and cities, a fact related to the low levels of female literacy and education prevalent, particularly in the 26 least developed countries (LLDCs) of Tropical Africa, where levels of contraceptive use are very low (Page, 1988). Only in the towns of Tunisia, Egypt, South Africa, Zimbabwe, Botswana and Mauritius do more than 40% of married women use contraceptives. Generally, family planning programmes have not been strong enough to have a major impact upon fertility. Traditional practices still play important roles: sexual abstinence before weaning; prolonged breast feeding; polygamy; child fostering; male labour migration. To these we must add the high incidence of venereal disease and latterly of AIDS, which are having drastic impacts upon city populations, especially in East Africa, where sexual activity is subject to less constraints than in the West and involves complex networking (Caldwell *et al.*, 1989).

Urban mortality

Africa has the unenviable reputation of having higher mortality than any other major world region. In 1990 its average life expectancy is only 53 years compared with an LDC average of 61 and an MDC average of 74. In many of the LLDCs it is less than 50. This is in spite of some general improvement since the 1960s through limited economic development, increased education and infrastructures along with some beneficial effect of public health interventions eliminating or containing major diseases like smallpox, sleeping sickness, yellow fever, yaws, leprosy, onchocerciasis and even malaria (Hill & Hill, 1988). Gains have been fewer in the LLDCs of the Sahel, the Horn of East Africa, Uganda and Mozambique, which have been devastated by disasters of both environmental and human origin, the latest of which is AIDS. These countries have some of the highest infant mortality rates in the world.

There seems to be no close inverse relationship between levels of urbanisation and mortality (Fargues, 1988), there being countries with relatively high levels of both (e.g. Equatorial Guinea, Senegal) and lower levels of both (e.g. Kenya, Zimbabwe). The fact is that patterns of change in mortality have varied greatly between countries. Nevertheless, it is the case that 'the children of women living in urban areas generally have lower child mortality than those of women living in rural areas' (Hill & Hill, 1988, p. 74), and the lowering of infant mortality in towns is even more striking (Fargues, 1988, p. 188). On the other hand, the independent effect of urban residence is small, other differentials such as parental (especially maternal) education, age of mother, income of household, birth order and region of residence being significant. The most obvious urban effect has been through the concentration of medical and pharmaceutical services, greatly reducing common infant deaths caused by umbilical tetanus, premature birth and enteritis.

Nevertheless, the picture of lower mortality in cities is not everywhere the same; for example, mortality in Nairobi and Mombasa has been affected by the congregation of so many poor rural–urban migrants living in awful housing conditions. The spread of AIDS, particularly associated with urban prostitution, will undoubtedly flatten out rural–urban mortality differentials in the future. Such differentials may also be reduced by governmental efforts to spread health spending more evenly over their countries in an attempt to provide a minimum level of health care to all citizens. Inevitably this means reducing allocations to large city hospitals in order to increase supplies of vaccines, antibiotics and oral rehydration therapy to rural areas.

Problems of city management of rural–urban migrants

Large African cities pose severe problems of management, which reflect not only rapid population growth but also: (1) the restrictions of original sites often selected for defence (e.g. islands, peninsulas, steep slopes); (2) complex ethnic and demographic structures associated with rapid growth; (3) marked areal differentiation and development of distinct quarters; and (4) inadequate economic bases. We only have space to highlight one or two of the management problems associated with a profusion of rural–urban migrants.

High rates of unemployment and underemployment are among the most intractable problems, although these terms are not easily defined given the widespread occurrence of the informal sector. Fuelled by rural–urban migration and natural increase, the labour supply of African cities far outstrips the demand. The slow growth of the modern sector never absorbs the labour supply especially of young adults, women and the newly educated. Underemployment in part-time work and in the informal sector has expanded massively in recent years, and it may account for 20–95% of the labour force in African cities (UNECA, 1989, p. 32), even though many may well be worse off than they were in rural areas (Gugler, 1976). However, this is neither universal nor easily determined in view of the frequency of multi-local residence and transient movements between town and country in Africa.

Poverty is so common in African cities and the wealthier elements so few that it is often difficult to accept that incomes are generally higher in urban than in rural areas and that much rural–urban migration responds to this differential, often as a supplement to rural incomes. The survival of the extended family and the strength of kinship ties is seen in ethnic differentiation and segregation producing polynuclear cities, but no doubt the continued growth of larger cities along with increased aggregate travel distance and costs of rural–urban migrants will reduce rural–urban links in the long run (O'Connor, 1983).

Housing difficulties in African cities arise from poverty and rapid influx from rural areas. Inner city slums are occupied particularly by people wanting short-term rents. More permanent sprawling spontaneous settlements, as around Kinshasa and Dakar, are not so common as around South American mega-cities. Most African countries have inadequate funds to cope with the swelling numbers in cities, but their urban-oriented policies are contributing to these numbers. Even if there were equitable regional allocations of funds, and this is rare, they cannot provide sufficient housing or public utilities (electricity, water, sewerage, refuse disposal). The latter are generally diminishing in adequacy as population grows, and thus the problems of pollution and environmental degradation accelerate, as seen in major cities like Lagos, Kinshasa and Addis Ababa.

Primate cities: generative or parasitic?

There has been a great deal of discussion over the years about the advantages and disadvantages of polarisation in primate cities, whether they are generative or parasitic, without there being any clear conclusions. This is partly because authors have tried to over-generalise: that all primate cities are either generative or parasitic, or that they are generative or parasitic from all viewpoints. Simple statements rarely reflect the complex reality of diverse African city growth, composition, functions and morphology, all of which are ever-changing. Thus, while colonial capitals such as Bangui, Kampala and Freetown go on growing at the expense of other towns and cities in the Central African Republic, Uganda and Sierra Leone respectively, Brazzaville (Congo), Harare (Zimbabwe) and Lusaka (Zambia) are much less dominating. Moreover, some indigenous cities like Addis Ababa, Cairo and Lagos have assumed international significance. Others, like Greater Khartoum (The Three-Towns), have strongly differentiated components. However, O'Connor (1983) detects a trend towards convergence, with many major African cities becoming more similar in characteristics; the more Westernised cities are becoming more Africanised and *vice versa*.

Among the generative features are the strong interrelationships with rural hinterlands, a factor that has increased in some countries, especially in North Africa since decolonisation and departures of large numbers of colonists. Centralisation and polarisation also have economic advantages through larger markets, labour supply and international links. However, among the parasitic features are the localisation of alien influences perpetuating colonial gateway functions, the excessive attraction to migrants, unbalanced populations, and the social diseconomies already referred to.

Generally, the disadvantages of large city growth are seen by African governments to outweigh the advantages, and nearly all have stated their intentions to reduce the population growth of their primate cities and to encourage more balanced urban systems through the development of medium-sized and smaller cities. Recommendations 40 and 41 of the Kilimanjaro Programme of Action for African Population and Self Reliant Development (UNECA, 1984) summarise the aims as follows.

'40. Countries should seek to integrate into the overall development planning process a comprehensive urbanisation policy which aims, *inter alia*, at reducing the current high migration to capital cities and other large urban centres, developing regional medium-size towns and ensuring an effective economic inter-dependence between rural communities and urban centres.

41. Countries should review their development strategies and incorporate into these strategies programmes which will stem the current flow of young people from the rural areas to the urban centres and ensure better living conditions in the rural areas. Measures should also be taken to upgrade living conditions in slum areas of cities.'

Depolarisation is therefore an attractive alternative to polarisation, but how can it be achieved?

Depolarisation policies

Until the 1970s little direct action was taken to deal with unsatisfactory population distributions, urban hierarchies and internal patterns of migration, African governments being more concerned with social and economic development than with spatial patterns of population. However, governments influence spatial patterns as much indirectly through social and economic policies affecting the allocation of resources as by direct policies of population redistribution. For example, closing the gap between rural and urban incomes may help to reduce rural–urban migration. Minimum wage legislation has been seen in this light in Tanzania and Zimbabwe, as have attempts to guarantee crop prices to farmers.

Among a number of urban-orientated policies of population redistribution, only a few countries like Libya have been wealthy enough to adopt a successful accommodationist policy to improve housing and employment opportunities, although most countries have been indirectly accommodationist insofar as they have focused educational and curative public health expenditures in their major cities. Indeed, more common have been closed city policies, especially in colonial times when migrants to cities were often prevented from entering or were returned to origin, a situation which has persisted longer in South Africa, and where urban ethnic dispersal has occurred with the creation of separate black townships. Urban dispersal by dormitory and satellite town development has also taken place around Cairo and Dakar. More revolutionary in principle has been the creation of new capitals in Nigeria, Tanzania, Ivory Coast and Malawi, but in practice Abuja, Dodoma, Yamassoukro and Lilongwe have not yet effected major changes in population redistribution. Abuja could become another Canberra, but much more significant in Nigeria has been the growth pole strategy associated with its fragmented federal structure, in which all state capitals have experienced strong urban growth. Elsewhere in Africa, decentralisation and regional development policies have encouraged the growth of medium-sized and small cities, as in Algeria, Tunisia and Kenya, but less successfully; there is always a danger that these smaller cities are viewed as merely smaller versions of primate cities having inadequate links with their hinterlands. Most students of African urbanisation – and of urbanisation in other parts of the Third World (Rondinelli, 1988) – favour policies that bring about more dispersed national urban systems. Furthermore, in the ethnically diverse continent of Africa, it also means that polarisation does not just favour one particular ethnic group.

Most Africans still live in rural areas, where governments wish to effect developments to stem the tide of rural–urban migration. Programmes of land colonisation and resettlement, and redistribution of colonised lands, have been implemented in Algeria, Kenya and elsewhere, and have been supplemented by massive 'villagisation' programmes in Tanzania and Mozambique, relocating millions of people (Clarke & Kosinski, 1982), not always to their greater wellbeing. Large capital-intensive development schemes, as along the Nile in the Sudan, have often proved too alien, too drastic and too polarising (Clarke *et al.*, 1985). Consequently, many African governments have seen more hope in integrated rural development programmes, which involve a variety of activities to reduce rural–urban disparities: land reform, more agricultural credit and technical assistance, improved transportation and social services, and greater investment in public works. Unfortunately, programmes in diverse countries such as Tanzania, Kenya, Ghana, Cameroon and Ivory Coast have had only moderate impacts upon rural–urban migration (UNECA, 1989, p. 45). Generally, they have failed to stem the rural–urban flows (Oberai, 1988). Programmes that have involved the people themselves in decision-making, which have been bottom-up rather than top-down, have been more successful. The full participation of the people is essential.

Conclusion

Policies of depolarisation have been most effective where they have not been unidimensional but have reflected overall policies of decentralisation and regional development. They must encourage the development of all regional resources, and avoid excessive economic concentration, urban orientation and rural under-development (Hope, 1989). Unfortunately, Africa's political problems have encouraged strong centralisation in order to engender the identity of new nation states, and so there has often been a conflict between political, social and economic ideals. Many governments, not only in Africa, have been more concerned with limited aims and political survival than the holistic development of their countries. It would be unrealistic in the current demographic, political and economic situation to expect any really significant slowing down of polarisation on primate cities. All the signs point to the fact that they are already in crisis (Stern & White, 1989), and will get worse.

References

Caldwell, J. C. (1976). Towards a restatement of demographic transition theory. *Population and Development Review* **2**, 321–66.

Caldwell, J. C., Caldwell, P. & Quiggin, P. (1989). The social context of AIDS in sub-Saharan Africa. *Population and Development Review* **15**, 185–234.

Clarke, J. I. *et al.* (1975). *An Advanced Geography of Africa*. Amersham: Hulton.

Clarke, J. I., Khogali, M. & Kosinski, L. A. (eds) (1985). *Population and Development Projects in Africa*. Cambridge University Press.

Clarke, J. I. & Kosinski, L. A. (eds) (1982). *Redistribution of Population in Africa*. London: Heinemann.

Escallier, R. (1988). La croissance des populations urbaines en Afrique. Quelques éléments d'introduction. *Espace, Populations, Sociétés* **2**, 177–82.

Fargues, P. (1988). Urbanisation et transition démographique: quelles interrelations en Afrique? *Espace, Populations, Sociétés* **2**, 183–98.

Gugler, J. (1976). Migrating to urban centres of unemployment in tropical Africa. In *Internal Migration* (ed. A. H. Richmond & D. Kubat), pp. 184–284. Beverley Hills: Sage.

Hill, A. & Hill, K. (1988). Mortality in Africa: levels, trends, differentials and prospects. In *The State of African Demography* (ed. E. van de Walle, P. O. Ohadike & M. D. Sala-Diakanda), pp. 67–84. Liege: IUSSP.

Hope, K. R. (1989). Managing rapid urbanization in the Third World: some aspects of policy. *Genus* **45**, 21–36.

Oberai, A. S. (ed.) (1988). *Land Settlement Policies and Population Redistribution in Developing Countries*. New York: Praeger.

O'Connor, A. (1983). *The African City*. London: Hutchinson.

Page, H. (1988). Fertility and family planning in Africa. In *The State of African Demography* (ed. E. Van de Walle, P. O. Ohadike & M. D. Sala-Diakanda), pp. 29–46. Liege: IUSSP.

Rondinelli, D. A. (1988). Giant and secondary city growth in Africa. In *The Metropolis Era* (ed. M. Dogan & J. D. Kasarda, vol. 1, *A World of Giant Cities*, pp. 291–321. London: Sage.

Stern, R. E. & White, R. R. (1989). *African Cities in Crisis*. Boulder: Westview.

UN (1989). *Prospects of World Urbanization 1988*. New York: United Nations.

UNECA (1984). *Kilimanjaro Programme of Action for African Population and Self-Reliant Development*. Arusha: UNECA.

UNECA (1989). *Patterns, Causes and Consequences of Urbanization in Africa*. Addis Ababa: UNECA.

18 *Urbanisation in the Third World: health policy implications*

T. HARPHAM

Introduction

Urbanisation, the rapid growth of the urban poor and the health problems of these populations in the Third World have received increasing attention during the past five years. The objective of this paper is to raise some of the issues that need to be considered in urban health policy development. The paper uses a particular conceptual framework to analyse health policy development, that is, the three-stage model of

(1) problem identification;
(2) policy formulation;
(3) policy implementation.

It is argued that in terms of urban health policy the 'problem' has, to a large extent, already been identified and that it is difficult to justify more research in this area alone. Various trends can be identified in the way urban health policy is being formulated, and two examples are provided. A number of constraints in policy implementation are discussed, and future research needs are identified.

Problem identification

Two parallel movements, which have recognised the needs of the urban poor in developing countries, have emerged in two distinct sectors at different times during the late 1970s and the 1980s. The first was in the housing or urban development sector; the second was the health sector. The second movement reflected the first: policy developments in urban health for the poor in the late 1980s are closely linked with an earlier policy development in the housing sector, namely slum upgrading or slum improvement.

Both of the above policy developments were partly a response to the sheer actual numbers and expected numbers of poor people living in intermediate and large cities of the world. The problems are acute, particularly in developing countries: estimates are that, at present, an

274

average of 50% of the world's population live at the level of extreme poverty, with this figure rising as high as 70% in some cities. On current estimates this means that by the year 2000 over one billion people will be counted among the urban poor (Harpham *et al.*, 1988). The detailed demographic trends and figures showing the rapid urbanisation that began in the late 1970s and continues today will not be presented here, as there are numerous summaries and good books on the phenomenon (see, for example, Gugler, 1988; Hardoy & Satterthwaite, 1986; Drakakis-Smith, 1987; and chapter 2 of Harpham *et al.*, 1988).

One of the best accounts of urban slum projects is given by Skinner *et al.* (1987). They document how, by the mid-1970s, many governments of developing countries had accepted that they would be unable to meet the housing needs of their low-income populations through government subsidised 'low-cost' public housing, and they would not keep 'squatters' or 'shanty town' residents away by bulldozing, or clearing their homes. This is not to suggest that these policies no longer exist: the most extensive and notable example of slum clearance in recent years was in South Korea for the Seoul Olympic Games in 1988. However, in many countries there emerged new policy directions in shelter delivery that is, 'slum upgrading' and 'sites and services' approaches. In both, the policy of supplying completed housing units was abandoned. 'Sites and services' introduced the construction of unfinished units and in 'upgrading', demolition, which only makes the housing deficit more acute, was reduced or halted and instead, improvement of existing substandard housing, slum upgrading, was undertaken.

At the same time that this shift in national government policy was taking place there were parallel developments in the non-governmental sector. This is well documented by Turner (1988), who presents a series of case studies from non-governmental organisations from developing countries.

By the early 1980s there were numerous slum upgrading projects under way in major cities of the developing world, for example Jakarta, Lusaka, La Paz, Guayaquil, Kingston and Madras. The emphasis of these projects was physical upgrading. For example, in the Kampung Improvement Project, which was initially a slum upgrading project in the major cities of Indonesia, 45% of expenditure was on roads (yet only 2.5% of the kampung, or slum, residents had vehicles) (Karamoy, 1984). An evaluation of this project in 1984 indicated that, although the physical environment had changed, there had been no significant change in the social or economic structure of the community. Similar conclusions were being reached in other slum upgrading projects throughout the world. This led to a change in policy. The Indonesian Kampung Improvement Project became 'integrated' in 1984. This essentially meant that the project became inter-sectoral and, for example, primary health care posts and primary schools were added to the previous road, drain, path, water and sanitation improvements.

Although the timing and details vary between countries, it is possible to discern a distinct policy trend in the early 1980s: a shift from single sector slum upgrading to multi-sectoral slum *improvement* projects (this shift in terminology appears to stem from the fact that 'upgrading' is perceived by many to indicate physical improvements only). So this was an opportunity for the health sector to reach the urban poor effectively. Were the health professionals ready for this development? How much of a priority were the urban poor? What had been happening in terms of 'urban health' while this development in the housing or urban development sector was occurring?

In the six years following the Alma Atma declaration on primary health care in 1978 the vision of primary health care and associated policy developments and literature was distinctly rural. This was a natural focus for policy debate as the majority of developing country populations lived in rural areas, and urban areas were perceived as a homogeneous over-served elite absorbing far too much of the national health budget. A seminal paper published by the World Health Organisation (WHO) and the United Nations Children's Fund (UNICEF) questioned this perception. The author, Alessandro Rossi-Espagnet (1984), used two arguments to promote urban primary health care. The first point was the rapid urbanisation of the developing countries. The second argument pointed to the rapid growth in numbers of the urban poor – residents of slums, shanty towns and squatter settlements – and the inequity that this created within cities (Rossi-Espagnet, 1984). This well-argued and well-disseminated paper stimulated organisations to support preliminary research, which explored the initiatives that existed in urban health care and examined the nature of urban health problems. For example OXFAM U.K., UNICEF and the U.K. Overseas Development Administration funded the London School of Hygiene and Tropical Medicine to hold the first international workshop on urban primary health care, bringing together urban health field workers from five continents for a workshop on *'Community Health and the Urban Poor'* in Oxford in 1985. In this period the active groups in innovative urban health initiatives tended to be non-governmental organisations (NGOs) and this is reflected in the book based on the Oxford workshop (Harpham *et al.*, 1988), where most of the case studies are NGO projects. The Oxford workshop identified several lacunae in the field, the most prominent being the lack of data that documented intra-urban differentials in health in developing countries. However, where studies did exist, they identified a grave pattern of high mortality and morbidity in poor squatter areas. For example in Porto Alegre, Brazil, the infant mortality rate in squatter settlements was 75 per thousand compared to 25 per thousand in non-squatter areas (Guimaraes & Fischmann, 1985). There was a call to move away from misleading aggregate city health statistics and to identify more clearly the health problems of the urban poor.

The late 1980s witnessed a burgeoning number of conferences, workshops and published articles, which focused on the health of the urban poor in developing countries. More light was shed upon the health problems of the urban poor but still few of the analyses focused on action. The coverage of countries was comprehensive. ORSTOM, the French overseas development research wing, held a workshop in 1986 in Dakar, Senegal, which covered Francophone African countries, with a resulting book (Salem & Jeanée, 1989). WHO and UNICEF continued to hold regional workshops in Asia and Latin America. These latter meetings resulted in the publication of the second book on urban health in developing countries (Tabibzadeh *et al.*, 1989). The issues of urban health had shifted from being 'in the shadow' to being 'in the spotlight' in a matter of five years. The problem had been identified. But how did this awareness of urban health problems reach and affect national governments and municipalities? Urban health was now on the agenda of many multilateral and bilateral agencies and non-governmental organisations. Was this interest reflected by urban health policy formulation within countries?

Policy formulation

The countries that are beginning to address urban health issues in a multi-sectoral way tend to be in Asia, where urbanisation is at a more advanced stage than Africa is yet a 'newer' phenomenon than in Latin America. Two different approaches for strengthening urban primary health care, as represented in Fig. 18.1, are emerging, often simultaneously within one country. The first approach is a sectoral one where the Ministry of Health attempts to strengthen urban primary health care (or often just health services) within the whole city in an unintegrated manner. It is often hoped that this 'strengthened' service will eventually have a trickle-down impact on the health of the poor. This strengthening may or may not be implemented through the municipal health office. The second approach is where the municipality acts as an umbrella organisation to coordinate and implement an intersectoral slum improvement project, which serves the poor communities only. These projects are usually administered by the ministry of urban development (or equivalent). An example of a country where these two approaches are developing simultaneously is India. Both approaches started in the early 1980s; at this time the Ministry of Health formed a high profile working group to make recommendations on how to

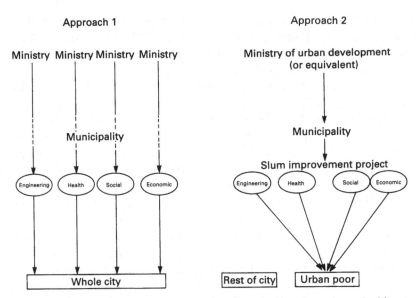

Fig. 18.1 Two different approaches to strengthening urban primary health care.

strengthen primary health care in urban areas (the 'Krishnan Committee'). One of the most important recommendations of the committee was to establish a health post for each slum, to be located in the slum itself, the strength of medical, paramedical, non-medical staff and voluntary health workers depending upon the population of the area. It seems that most Indian states failed to respond to these recommendations, partly because other components of an effective delivery system like training, supervision, monitoring and evaluation, supplies and information, education and communication activities, were not addressed (and therefore not funded) by the central Government of India (National Institute of Health and Family Welfare, 1989). As a response to this poor uptake of their recommendations the Ministry of Health is, in 1990, undertaking a needs assessment and a review of the existing status of service delivery, with a view to another future attempt at 'strengthening' traditional ministry-led health service delivery to urban populations.

It is worth noting that although few municipalities have taken note of the Krishnan Committee's recommendations, Greater Bombay is an exception. Here a number of 'health posts' have been established in poor urban areas. Yesudian (1988; reviewed in Harpham, 1988) undertook a household survey around one of the newly created posts in the Bombay slums and concluded that residents knew about the existence of a health post but did not understand its function, basically because no outreach work was undertaken from the post to contact residents. It was found that one of the main obstacles was that health posts are located in existing dispensaries, maternity homes or hospitals and that health post staff often are diverted into other work such as immunisation camps. Like many other studies of health facility use by the urban poor, this study found that community members were still using the general hospital and medical college hospital for treatment of minor ailments, despite the supposedly 'strengthened' primary health services. This may imply that relying on traditional (even if strengthened) health delivery is not enough. What are the alternatives?

At the same time as the Indian Ministry of Health was attempting to strengthen urban primary health care, the Ministry of Urban Development began to support a number of slum improvement projects that included a substantial health component. The projects are intersectoral with coordination provided by the municipality. The Indian Ministry of Health has no control or particular input into these projects. For example, when designing their projects, some of the Indian cities implementing slum improvement chose to bypass norms laid down by the Ministry of Health (e.g. number of households covered by a 'Voluntary Health Worker', the Indian version of a Community Health Volunteer). It is worth noting that the Indian health ministry staffing norms tend to have been developed for rural settings and may not be appropriate in the urban milieu. The same has been true for norms of other cadres working in the urban environment. For example, the 'community organisers' of Indian slum improvement projects are a particularly urban phenomenon and norms for them have been specifically for the urban context. In urban projects, staff may be 'project'-adapted for urban slum programmes, but do not fit in with national, ministry-organised staff norms. This can allow flexibility within the project context, but when the project is large (and consequently has many staff) this has long-term implications for local governments in knowing how to coordinate idiosyncratic project staff with traditional cadres in ministries or other local government settings.

Slum improvement projects are also emerging in other countries. They tend to be neat, time-bound, targeted packages of action, which have attracted donor attention and support because of their direct poverty focus. However, this can create problems of sustainability for national and local urban planners, who continue to run the rest of their city in stratified sectors.

The two approaches in urban health management that are emerging do not rest easily together. In response to a proposal for a health survey to help design the health inputs of the Jakarta Slum Improvement Project, a senior member of the Indonesian Ministry of Health requested that a separate survey be done to serve the needs of the MOH in strengthening urban health. The 'clients' and the topics addressed were to be the same. Such territorial rivalry is just one factor to be overcome in the process of developing urban primary health care. As more projects and initiatives emerge, more key issues of implementation can be identified. Some of these are considered below.

Policy implementation

Many projects that have begun to implement primary health care in poor urban communities in developing countries have experienced common problems. A summary of a selection of these problems is presented in Table 18.1. Many of these problems, or constraints, are characteristics of an urban environment and would not arise in a rural context in developing countries: for example, the dependence upon hospital services; the dependence upon cash economy; the lack of land on which to grow food.

More research is needed on the implementation of urban primary health care. There are very few data on the costs or alternative methods of financing urban primary health care, for example. The issue of sustainability needs longitudinal research, and the management capacity of municipalities to take on board urban primary health care needs assessing. These are just some of the issues of implementation that warrant research in the future.

Conclusions

The health problems of the urban poor have largely been identified, and one could argue that there is no need for more studies to measure the extent of the problem (in terms of describing the morbidity, mortality and nutritional status of the urban poor). Policy formulation is currently under way in a variety of arenas: multilateral agencies, bilateral agencies, non-governmental organisations and national governments. Research into how and why urban health policy is formed would provide insights into the process, but above all research is needed on the *implementation* of policy. Which approaches are most cost-effective, sustainable and manageable, and involve the community? These are the questions for urban health researchers in the 1990s.

References

Drakakis-Smith, D. (1987). *The Third World City*. London: Methuen.
Gugler, J. (1988). *The Urbanization of the Third World*. Oxford University Press.
Guimaraes, J. J. & Fischmann, A. (1985). Inequalities in 1980 infant mortality

Table 18.1. *Constraints to environmental improvements and the provision of community health services in poor urban areas of developing countries*

Constraint	Potential solution(s)
1. Established dependence on curative service (e.g. private practitioners and hospitals) makes it difficult for PHC[a] to achieve credibility.	Begin project with affordable, accessible curative services; add preventive activities gradually. Focus on rationalising referral system.
2. Heterogeneous groups make definition of 'community' difficult.	Social analysis of neighbourhoods before definining 'communities'.
3. Greatest perceived need often security of tenure and improvements to housing.	Obtaining security of tenure as priority. Self-help housing schemes complement environmental health improvements.
4. Dependence on cash economy so difficult to pay CHWs[a] 'in kind' as often occurs in rural areas.	True volunteer CHWs *or* avoid CHWs *or* municipality employs the CHWs. The trend is towards the latter.
5. Little land on which to grow food to improve diet and nutritional status.	Income generation through skills training, facilitating low-cost loans.
6. Difficulty of coordinating sectors.	Use the municipality as umbrella organisation; form an urban community development department within municipality.
7. Coordinating small-scale NGO[a] activities difficult.	Form health coordinating committees at community and/or municipal levels.
8. Maintenance of physical assets (e.g. roads) difficult.	Form maintenance agreements between community and municipality.

[a] CHW, community health worker; PHC, primary health care; NGO, non-governmental organisation.

among shantytown residents in the municipality of Porto Alegre, Rio Grande do Sul, Brazil. *Bulletin of the Pan American Health Organization* **19**, 235–51.

Harpham, T. (1988). A review of Yesudian's study of health posts in Bombay. *Health Policy and Planning* **4**(4), 368–9.

Harpham, T., Lusty, T. & Vaughan, P. J. (eds) (1988). In *The Shadow of the City: Community Health and the Urban Poor*. Oxford University Press.

Hardoy, J. E. & Satterthwaite, D. (1986). *Small and Intermediate Urban Centres*. London: Hodder & Stoughton.

Karamoy, A. (1984). The Kampung Improvement Programme: hope and reality. *Prisma (The Indonesian Indicator)* **32**, 19–36.

National Institute of Health and Family Welfare (India) (1989). *Report on assessment of family welfare/primary health care needs in urban areas (specially slums) in cities with population more than two lakhs and formulation of proposals for their strengthening* (monograph). Government of India.

Rossi-Espagnet, A. (1984). *Primary Health Care in Urban Areas: Reaching the Urban Poor in Developing Countries*. A state of the art report by UNICEF and WHO. Report No 2499M. Geneva, Switzerland: World Health Organisation.

282   T. Harpham

Salem, G. & Jeanée, E. (eds) (1989). *Urbanisation et Santé dans le Tiers Monde.* Paris: Orstom.
Skinner, R., Taylor, J. L. & Wegelin, E. A. (eds) (1987). *Shelter Upgrading for the Urban Poor: Evaluation of Third World Experience.* Manila, Philippines: Island Publishing House.
Tabibzadeh, I., Rossi-Espagnet, A. & Maxwell, R. (1989). *Spotlight on the Cities: Improving Health in Developing Countries.* Geneva, Switzerland: World Health Organisation.
Turner, B. (ed.) (1988). *Building Community. A Third World Case Book.* London: Building Community Books.
Yesudian, C. A. K. (1988). *Utilisation of health services by the urban poor (a study of the Naigaum maternity home health post area). A study conducted by the Tata Institute of Social Sciences, Bombay and sponsored by the World Health Organisation* (monograph). Bombay, India: Tata Institute of Social Sciences.

Index

Printed in the United States
By Bookmasters